Compass Companions

A Collection of Canadian Travel Stories

Jeannie Henke

Unforgettable travel stories that will fire your imagination,
lift your spirits and rekindle travel camaraderie
for the Canadian traveller's journey around the world.

Published by
Compass Companions
Silver Springs P.O. Box 71094
Calgary, Alberta
Canada T3B 5K2
www.compasscompanions.ca

National Library of Canada Cataloguing in Publication

Henke, Jeannie
Compass Companions: A Collection of Canadian Travel Stories/Jeannie Henke.
Includes index.
ISBN – 0-9780160-0-9

Editing: Jeannie Henke, Susan Smith, Ian Tennant, and Jen Gouthro

Printed in Canada by
Houghton Boston, Saskatoon

This book is dedicated to my children
Justin and Johnathan
and my husband Larry

In appreciation to my family and friends, especially
Ashifa Lalani, Louisa Powell, Laurie Benz,
Cheryl Geeraert, Tina Hansen, Evelyn Bowie
& Mr. Anthony Burley, my Grade 9 teacher

With gratitude to the many Canadians
I encountered while sharing their stories.
That alone was a privilege and an adventure.

For the traveller in all of us!

TABLE OF CONTENTS

INTRODUCTION

We have embarked on an enormous coverage of Canadian travel from representatives within the Canadian travel industry. This includes stories from travel editors and writers, educators of tourism, tour operators, travel agents, tourist boards, pilots, and the all-adventurous Canadian! Follow each story from beginning to end as you are taken around the world covering the seven continents. Each unique story allows you to be immersed with your fellow Canadian's amazing travel experience from the comfort of your home. Share in their enthusiasm as Canadians discover a hidden corner of Canada, mingle with diverse cultures or navigate the open seas.

Compass Companions reveals the best of the spirit and character that exemplifies true Canadian adventurism. So sit back, relax and find out what your fellow Canadian companions are doing around the world.

Jeannie Henke, BA

Introductory to the High Granite Wilderness: A Canadian Bush Pilot

John S. Goulet

One of the great pleasures of being a bush pilot is being able to enjoy the wilderness without suffering the extreme kind of pain and hardship that trappers, prospectors, miners, and drill crews must endure. The bush pilot can make a short two hour flight into one of these remote wilderness camps, enjoy the occupant's enthusiastic company, indulge in the freshly baked bannock hospitality, and wing his way back to civilization and his own warm bed for the night.

This lifestyle, of course, does not apply to helicopter pilots or pilots on contracts where the aircraft must stay with the geological or drilling party in the high arctic, but it does apply to the majority of the bush flying jobs I have had.

The most important rule I learned was to generally take care of the camps as if they were your own, and specifically to be good to the cook or the camp boss or whoever was in charge of the kitchen. There is nothing better than a cup of hot coffee and a pile of fresh pancakes (made from scratch) topped with cool yellow butter and real maple syrup, and a side of crisp Canadian bacon to fuel the pilot on his return flight south.

This story, taken out of Grant McConachie's (former President of Canadian Pacific Airlines) biography *Bush Pilot With a Briefcase*, epitomizes, to the extreme, an example of what it is like to be the bush pilot with a responsibility to his trusting customers left in some remote wilderness. Hopefully, we will never get to experience what it feels like to be the stranded fool waiting faithfully on the pilot to return.

I can't help but remember the time I left an Indian trapper and his family on a remote lake out from my base of Island Lake, Manitoba. I loaded the trapper, his son and daughter and wife, an outboard motor, gas can, backpack, and rifle into the C185 floatplane. By the size of the pack I could see they carried only the staples of flour, salt, and lard. We flew for about one hour as he directed me to his remote camp. I dropped them off and agreed to pick them up in two weeks.

The next two weeks was filled with ten hour flying days, dispersed with fights with my employer, inquiries by the RCMP, distressed nurses, an Indian boy with a fishing hook through his finger, burning houses, and other life-on-the-reserve distractions. The two weeks turned to three when suddenly I remembered my promise. I set out the next day only to discover to my horror that I had not marked the location on the map. With three weeks of ten-hour flying days, I had been everywhere imaginable and could not even remember what lake I had dropped them off on.

I flew to where I thought was the location and started a slow search pattern. As luck would have it, I spotted a fire on the horizon after about half an hour of searching. I circled and to my relief found the family standing on the Precambrian rock shoreline.

Courtesy of Canadian Tourism Commission

Gunisad Lake Lodge, Manitoba

I landed and nosed the floats to the shore. We loaded the aircraft without a word spoken. No one seemed perturbed or upset, and in fact, they did not even seem to notice that I was late. They had shot a small moose so we loaded the hide, the dried meat, some fresh meat they had buried, and the head, and flew back to the Island Lake settlement.

Only days later did I start to get the friendly polite ribbing from the Indians in the settlement about how some "new" pilot had forgotten the Spencer family in the bush and had only found them by accident when he was wandering around lost one day. The smile induced wrinkles in the corners of their eyes told me they knew damn well who the pilot was. I enjoyed the way they dispensed this good humour teasingly, but I also recognized they were doing it to let me know that, as their guardian of transport, I too had responsibilities to this god-forsaken community.

John is the editor of Virtual Horizons.
www.bushpilot.com

Nova Scotia Cooking:
The Way to a Traveller's Heart

Nova Scotia Tourism, Culture & Heritage

If you're from "away" you're probably aware that Nova Scotia serves up some of the best seafood in all of the Eastern Seaboard. However, what you may not know is that the province's regional cuisines are equally delicious.

For a province full of good, and some would even say great cooks, the locals are modest about their home-grown tasty dishes enjoyed by some lucky visitors, often declaring that it's just "plain cooking."

Nova Scotians will tell you that it's the way they were brought up with fish in one form or another every Friday; baked beans and brown bread on Saturday nights and Sunday mornings; fish chowder for supper (yes, supper, not dinner); fish cakes at least once a week, always served with homemade chow; spareribs and sauerkraut; boiled dinners of corned beef and cabbage, followed by leftover hash; and not forgetting, delicious rabbit stew, which you might have trouble finding on today's menus. But, Nova Scotians call these local culinary delights everyday fare!

Go to any restaurant worth its salt in the capital city of Halifax or across the province and you'll find chefs and cooks, often hailing from "away," who revere what Nova Scotians consider to be nothing out of the ordinary. Loved, yes, but what's so different about it? In most homes, cooking is considered routine, the only love being in the eating. Sometimes, it takes an objective outside view to fully appreciate that lovingly prepared local ingredients can mead into scrumptious meals that are definitely worth eating.

As early as 1633, when Acadians came to stay, they planted their apple trees and, urged on by their frugality, they invented "receipts" using indigenous ingredients, firmly establishing their rich culinary heritage. Other local inhabitants with French, British (especially Scottish), Loyalist, and German backgrounds also influenced the unique cuisine that emerges from today's Nova Scotia kitchens.

Some say the love of the traditional Acadian "rappie" pie is acquired only after a few helpings. Others take to it immediately, especially when it's served with molasses, which is never far from the kitchen table. If that doesn't suit your taste, you can be sure a chicken fricot or clam pie will do the trick. And for dessert, you can't do better than buttermilk pie, an ingenious use of the by-product of butter making. To leave Nova Scotia without having tried some form of Acadian fare is to have never really been there!

As a visitor, you may not find yourself sitting at a typical domestic kitchen table in Yarmouth, Guysborough, or Hants County. But if you do, the person doing the cooking might fancy it up by using a recipe from a trendy magazine in an effort to please you. However, you can buy the same magazine, and probably have. So, don't be shy! Let your host or hostess know that what you crave is that special flavour that comes with authentic Nova Scotian cooking. You may have heard about the wonderful early summer meal of new

Picture Courtesy of Nova Scotia Tourism, Culture & Heritage

vegetables cooked in cream (or milk) called Hodge Podge. Try it and judge for yourself how plain olè, well actually, new vegetables can become something ethereal, which they do!

Nowhere on earth is the strawberry season more fully enjoyed than in Nova Scotia, especially when the sweet berry is married with the tart rhubarb. If it's on the menu, order a piece of strawberry-rhubarb pie. It's one of the local favorites. And in August, when the blueberries are being picked, you'll want to sample that dish called Blueberry Grunt. Forget that it's only blueberries simmered with sugar and spices and topped with dumplings! Dumplings you say? Together they're great! Look for it and ask for it.

Perhaps nowhere is the English culinary tradition more prominently reflected than in the pubs and taverns across Nova Scotia. This is where you'll find the shepherds' pies, plowman's lunches, sausage rolls, and most popular of all, fish and chips, accompanied by a pint of one of Nova Scotia's popular, internationally-recognized brands of brew.

Fish and chips were to industrial England what the hot dog was to the United States, the original fast food of generations who knew of McDonald's only as a Scottish clan. And here in Nova Scotia, there are many corner operations and small restaurants that proudly aim to please the fish and chip crowd. Upscale restaurants might call their dish haddock and frites, but locals know its fish and chips and order it at places they have come to rely on.

Food critics have tried to locate the best of the batch, but it all comes down to a matter of preference. Do you like your fish fried in light or heavy batter? Or just run through a coating of seasoned flour? And will that be cod or haddock? Some ask only one question before they order, "Is the fish fresh or frozen?" If the answer is evasive, like "Well, it's fresh-frozen," true fish and chip addicts opt for the special of the day, which is usually fresh off the docks.

Lunenburg County is the stronghold of German fare. If you're lucky, you might start with an appetizer of Solomon Gundy (pickled herring) and go on to a salad of sauerkraut or cucumbers in sour cream. Follow this with an entrée of salt cod and pork scraps, called "House Bankin" by some or "Dutch Mess" by others, and for dessert, gingerbread with lemon sauce, or pie of any kind. It's here that the great cabbages grow and are turned into sauerkraut and kohl slaw. It's here where potatoes are made into a delicious "Kartofflesuppe" or added to the batter of raised doughnuts called "Fasnaks". It's also here that a sausage called Lunenburg Pudding is adored, and where some of the best home cooks ply their mixing spoons and rolling pins. Oh, yes, in Lunenburg County, they take pride in their baking but modestly ask, "What's all the fuss?"

You'll find Scottish cuisine prominently located in Pictou County and Cape Breton, or wherever oatcakes appear in breadbaskets (give a restaurant a star for this). From Scotch broth or "cock-a-leekie" soup served with bannock or porridge bread, to smoked kippers, finnan haddle or poached salmon with a side of clapshot, or tatties and 'neeps', you'll find something special about Scottish fare. This is especially true if you wash it down with a tot of rare single malt whisky distilled at Cape Breton's Glenora Distillery, the only place in North America that makes it.

With the early back-and-forth flow of people between Nova Scotia and New England, it is sometimes difficult to distinguish between the cuisines of the two regions. However, the locals readily credit New Englanders for transplanting to their Nova Scotia homes, the cornmeal that is turned into today's puddings and Johnny Cakes; baked beans and brown bread; and the traditional Thanksgiving dinner of turkey, cranberry sauce and pumpkin pie.

Nova Scotians hold close to their hearts the claim to the first fish chowders ever served in North America. To stake their claim, they point to the blackened pots of the early 1500s, when foreign fishermen came to fill their nets and ships with the prolific cod. They had to eat and local fish were tasty and plentiful.

Today's visitors will surely find the perfect fish, seafood or clam chowder somewhere in Nova Scotia's cities, towns or hamlets. But finding it depends on our personal preference. Some connoisseurs like a thick chowder, others like it thin. Perhaps somewhere in between lies perfection. Your version of the perfect, heavenly chowder awaits you, if you stay long enough to savour the culinary and numerous other delights of Nova Scotia, Canada's Ocean Playground.

Ireland:
Exploring the Rugged West Coast

Jen Gouthro, Editor & Writer

We could not have been more pleased. Our rental car turned out to be a cheerful kelly green Fiat, a fitting vehicle in which to explore the beautiful Dingle Peninsula on Ireland's rugged west coast.

Jen & friend at Dingle Peninsula.

The Dingle Peninsula is the westernmost peninsula in Ireland, and the lesser-known cousin to the touristy Ring of Kerry. Its silver-sanded beaches and jagged cliffs have been backdrops for such movies as Ryan's Daughter and Far and Away, and it's easy to see why the Dingle Peninsula is growing increasingly popular as a tourist destination.

Our first stop on our Dingle tour was the charming town of Dingle, home to 1,500 inhabitants and more than 50 pubs! Dingle also serves as the residence for Fungi, a friendly dolphin who has been charming locals and tourists alike since the mid-eighties. Surprisingly, Dingle offers more than good traditional pub grub and a cold pint. You can pick from gourmet restaurants and trendy cafes if your cravings are more sophisticated. We opted for some homegrown craic (Gaelic for lively conversation) coupled with fish and chips at Murphy's Pub on Strand Street.

After debating which one of us was best equipped to drive the 30 plus miles of treacherously thin, winding roads of Slea Head Drive (on the other side of the road, no less!), we departed for magnificent Slea Head. Despite Dingle being a Gaeltacht (Gaelic-speaking) region, the locals are kind enough to have signs written in English and Irish. On the way, we stopped to snap some photos of the awe-inspiring emerald green hills, dotted with clans of white sheep. Any plans to make good time around the peninsula were forgotten as we rested awhile amidst the breathtaking beauty. We quickly realized that renting a vehicle and taking our time to explore this region was a great idea.

Can you really call it a road-trip without getting lost at least once? A few wrong turns apparently made a right when we happened upon a rocky inlet protecting what we realized later was popular Ventry Beach. We trekked down the steep path and gasped at the tenacity of several swimmers braving the still chilly June waters. We couldn't decide which had the deeper blue hue – the cloudless sky or the unrelenting waves, so we declared it a tie.

It was my turn to take the wheel. I was white-knuckled as impatient locals sped by me on the sharp turns. Undaunted, I took my time so I could enjoy the lovely drive. Nonetheless I was relieved to spot the normally cloud-obscured Blasket Islands in the distance, confirming our arrival in breathtaking Slea Head. Sharp outcroppings of rock lined the base of the craggy cliffs and green waves exploded into soaring arcs of foam as they reached land.

The spectacular shoreline was on one side of the road, and on the other side, magnificent stone fences interrupted steep, grassy hills. There were plenty of sheep to coo over as we explored Dunbeg Fort, a series of stone "beehive" huts reconstructed from Iron Age and Early Christian ruins. The Fort (or Cashel as it was also called) was occupied by five structures, one of which bore an inscribed cross. The huts were interesting, but the view from the hill was more than worth the price of admission.

My travel mate Lynne wanted to experience the hairpin turns of Dingle Peninsula, so we changed drivers again. As we travelled north past the village of Dunquin on the way to Ballyferriter, we noticed some cars parked alongside the road. We couldn't resist stopping to investigate, lest we miss a fantastic sight. We weren't disappointed.

As we walked up the hill, we were greeted not only by a majestic view of the Irish landscape, but we had musical accompaniment as well! A harpist played traditional tunes beside a box of quickly moving homemade CD's. We sat and listened to the soothing music and meditated upon this movie-perfect moment. It was wonderful to relax and drink in the view. Several sheep roamed about, indifferent to our presence.

We were successful in tackling Slea Head Drive in half a day, but were determined to explore North Dingle before pushing on to the town of Tralee. After a late lunch, we drove north from Dingle along a narrow Cliffside road via majestic Connor Pass, the highest mountain pass in Ireland. We were surprised to encounter our most challenging roads yet, with barely one small car being able to squeeze through in some places. I felt like a world-class driver after navigating this area! For me, Connor Pass was the most beautiful part of the trip.

The scenery was otherworldly: rocky, windswept hills emerged from fertile green valleys, with the basin floors punctuated by glassy glacial lakes. We felt fortunate we added this leg of the trip to our Dingle adventure. "Now, this is Ireland!" my friend Lynne announced. I agreed.

We soon realized our weariness and decided it best to depart for Tralee. The town was a convenient stopping point after a day on the Dingle Peninsula and had an abundance of comfortable accommodations and quaint eateries. We meandered through the surprisingly bustling streets until we happened upon the Abbey Inn, former stomping grounds for many Irish bands, including U2. We savored the atmosphere as well as our cold pints, and reflected upon a day jam-packed with sand beaches, jagged cliffs and true Irish hospitality.

Sl`ante! (Gaelic for Cheers!)

A Wedding in Chelsea: Going to the Chapel of Love

Teresa Howe

Photo © Ritz London Hotel

Chelsea. During the sixties, it was a hotbed of fashion, rock 'n roll and cafes. In the early 16th century, both King Henry VIII and Sir Thomas Moore built their mansions in this riverside community. Over time such luminaries as Oscar Wilde, Rossetti and Whistler made it their home. Today, Chelsea, the royal borough, is still colourful and very much known for the famous Kings Road and its many boutiques, restaurants, theatres, celebrities, the Royal Hospital, the Physic Garden and the Albert Bridge, but in 2005, it became my home away from Toronto.

The move to London had been prompted by my husband Tom's executive transfer by his employer. While it was all very exciting for Tom to accept a challenging international insurance position for a period of two years, I had given up my Human Resources consulting business and became in charge of settling us into a new life. I was thrilled to be living in the hub of culture, theatre, history and royalty, and decided to embrace this time of adventure and also work on my health and fitness after many years of neglect. Thus, I joined a health club and was now religiously attending various exercise classes. It was also an efficient way of working off the delicious "puddings" (desserts) pub food and fabulous cheese courses that had become my addictions. That morning was no different than any other except that it was my wedding anniversary and I was anticipating bubbly champagne and a lovely dinner celebration with my husband that evening – even greater reasons to work extra hard at the class.

August 25th was a very warm morning as I came out of the Pilate's class, perspiring and exhausted, and stood on the corner of Kings Road next to the Chelsea Town Hall chatting to a friend. A young, well-dressed, attractive couple approached us, stopped, smiled and then looked to us with questioning faces. The young man just blurted out in a proper English manner, if we would be kind enough to be witnesses to their wedding and if we had half an hour to spare.

We looked at them and thought that perhaps they were joking. They insisted with a giggle. I said, "Yes, I can do it." Ewa, my friend, hastily said, she had to go. She glanced at me with a smirk and walked away. I immediately and apologetically said to the couple that I was not dressed appropriately, in my sweats after a workout, but they said "Not a problem!"

They then scouted around for another witness and approached a lady, Barbara, from my Pilate's class who, looked rather shocked and blurted out "Yes" and then added, "Well you know, I have been married for over fifty years and never been inside this registry so it will be a first for me."

As the four of us entered the Chelsea Town Registry for the service and separated to be briefed, we saw rows of empty chairs, which would easily accommodate fifty guests. The wallpaper was rich in golden brocade complemented by elegant jade-coloured carpeting. Enormous vases of oriental lilies with mollucella, golden chrysanthemums, flowing beargrass and rich green palm leaves filled the room with their sweet scents. It was an elegant and historic atmosphere fitting for an elegant wedding party. Barbara and I glanced at each other and shook our heads as we realized just how comical the entire situation was – two tired and frazzled strangers in workout clothing, and two young romantic sweethearts in all their finery. Barbara leaned over and whispered, "I understand that the famous Wallis Simpson married her second husband, Ernest Simpson right here in 1928."

Suddenly almost enchantingly, the beautiful wedding march from A Midsummer Night's Dream sounded as the groom appeared first, very handsome in a crisp navy suit, and striped navy and gold tie. His floppy black hair, that he nervously pushed back, reminded me of Hugh Grant's locks. A minute later, the bride followed down the empty aisle, wearing a wispy, cream-coloured dress with dancing cherry and mint flowers scattered throughout the satin and chiffon. In her shoulder length auburn hair, she wore an ornamental pearl comb, which nicely matched her pearl coloured single orchid clasped in her delicate, white hands.

The room felt magical as the amiable Commissioner welcomed the four of us and spoke beautifully and humorously about the meaning of marriage. His eyes twinkled when he asked us if we had any objections to this marriage just before we signed the register.

At the end of the ceremony, the bride asked us to take their photos but her camera would not cooperate and the groom had forgotten his, so once again we looked at the comical side of this day as the groom ran out of the hall, down the stairs to the street and over to Waitrose (a grocery store) to buy a dispensable one!

When Barbara and I were dismissed from our witness duties, we realized that we had not been formally introduced nor were we invited to a reception or for a glass of champagne. All we knew was that the couple had been living together for seven years in Cheswick and tomorrow were flying off to Malaysia for a blessing and then returning for a reception in London.

I wished them much happiness and told them that August 25th was also my wedding day and it must be destiny and good luck for them to have selected me to be part of their special day. I left with misty eyes and a romantic feeling.

Later that evening, when I recounted the story to Tom over our romantic anniversary dinner, his only comment was that this could only happen to me. I cannot help the feeling that somehow their story could be altogether different. I keep asking myself – did they not have any friends in London to witness their vows? Maybe they married against their parents' wishes? They were young enough – maybe there was an arranged marriage from which they ran away? Or maybe there is a broken heart somewhere along the way? Or, even worse still – I have become an accessory to a "crime" and the family will come after me? My imagination is running away with me.

This memorable event adds one more reason why my stay in London is such an enjoyable adventure. When we return home to Canada, I will have many stories to tell.

Afternoon Tea at the Ritz London

Thinly sliced cucumber, smoked salmon, egg mayonnaise & mustered cress sandwiches, followed by scones with homemade strawberry jam clotted cheese & a selection of pastries. Served with a variety of tea with cream or lemon slices.

The Ritz is absolutely fabulous, I did peak in! (There is a six-week waitlist for Afternoon Tea for non-guests.)

Mind Your Language!
Belgium

Linda Barghoorn

Moving to a new country often involves a myriad of adjustments and challenges in one's life – from culture to climate, passing a driving test in a foreign country to learning to drive on a different side of the road, making new friends and leaving old ones behind, and among the most difficult: learning a new language.

Our move to a small rural village on the outskirts of Brussels, Belgium, seemed a perfect opportunity to resurrect my rusty French and put it to use in an attempt to integrate into local Belgian life. We chose a home a few miles from the international school our children would attend – in a small, pretty little village near Napoleon's historic battlefields – where French was a "must" in order to communicate.

This was in marked contrast to the community ten minutes away, which hosted the international school. Many expat families chose to live in what was affectionately known as the 'American Ghetto' – the largely expatriate neighbourhood surrounding the school. Here, the shopkeepers were dramatically more bilingual in an effort to cater to the large numbers of English-speakers. This was not for me! I was determined to make our lives more fulfilled (a.k.a – difficult) by forcing us to immerse more fully in the native language!

My clever thinking was that by placing us in such a position, we would have no choice but to practice and expand our French language skill. The neighbours spoke French, the shopkeepers and restaurateurs spoke French, as did the staff at the sports complex up the road where we took a family membership. If we wanted to socialize, shop, eat out or work out it was going to involve the use of French. We were going local!

And this is a good thing, right? I pitched this to my kids as the perfect opportunity for them to learn a new language – one that was of utmost importance for us if we wanted them to be bilingual upon our eventual return to Canada. But, learning a foreign language is more than just an opportunity to add another language to your repertoire; it offers a glimpse into a new culture, way of expression and different attitudes. It's an opportunity to broaden your horizons and meet people who you might otherwise never encounter, never share a conversation or a laugh…

There were many laughs along the way as I struggled to dust-off my archaic French skills and dive in feet first. My first encounter with our neighbours a few houses down the road involved the attempted rescue of a deer, which had wandered into our fenced yard and then became disoriented and completely distraught at not being able to find the open gate by which it had entered. We introduced ourselves briefly and tried to communicate that we needed their help in corralling an animal from our yard before it entangled itself hopelessly in the fencing that it was trying to break through. But the word for "deer" escaped me. I tried 'binche' – to which they looked at me in complete confusion and replied 'vache?' ("Cow?"). Now, this seemed ridiculous to me - that they thought a cow had wandered into our yard - as there were no cows for several miles. I found myself at a loss; but luckily my husband came to my rescue, resorting to gestures, putting his hands to his head like antlers and telling them, "Bambi, Bambi!" Miraculously, this worked! They came to help. We rescued the deer.

I returned home to search for "binche" in my French dictionary and didn't find it. Like a light switch abruptly switched, it suddenly occurred to me that "binche" was the name of a Belgian city, famous for its Carnival celebrations. At this, the neighbours' suggestion that a cow had been in the yard seemed less ridiculous than what I'd offered up! But I still hadn't found the French word I'd needed. I flipped to the English-French section of the dictionary and looked up "deer." There it was: biche. The word is now etched permanently in my brain. If I ever need to rescue another deer in a French-speaking country, I'll be fine! I also marvel at the workings of the human brain; that somewhere buried in my grey matter was the word; and that I came so close, and yet so far at the same time!

'So close and yet so far', came to be the theme of many of my conversations with the locals. Sometimes the French word I needed was just buried too deep in my brain to be pulled out at a moment's notice. Other times I just lacked the vocabulary. In these instances I tried translating literally from English to French using words that I did know. Sometimes it worked; often it wasn't that simple.

I expected my first visit to the hair salon to be a straightforward event, but looking back I realize I probably should have made some advance preparations. It's hard enough to communicate what you want to a hairdresser – especially one you're visiting for the first time – in your own language, let alone a foreign one. But all I wanted was a trim, and so I thought, "How hard can it be?"

When asked by the handsome, flamboyant male hairdresser what was to be done that day, and realizing I didn't know the French word for a "trim," I tried the phrase "une petite coupe." In my mind "a little cut"... not a "big" haircut... should have gotten the idea across. While it seemed relatively obvious to me, the initial response by the hairdresser was a sly smile as he remarked rather tongue-in-cheek, "We could do that!" With a few more phrases and a couple of gestures, I was able to communicate what I wanted and managed a half hour later to exit the salon looking rather nice.

Despite the slight stumble, I was quite happy that I'd managed to communicate what I wanted and did not walk out with a ridiculous new coif because of some bizarre language miscommunication. But, the rather quizzical look on the hairdresser's face at my initial request, nagged at me, and a few days later I asked a French-speaking friend whether I'd said something wrong. She explained that a "petite coupe" could mean "a fling"... which certainly explained the rather bemused smile by my hairdresser friend. I've never pursued further whether this is actually a translation for my botched attempt... But, I do now know that a hair "trim" is a "coupe des points" ("cutting of the ends"). I've never run into that hairdresser again to take him up on his offer!

It might have been tempting to avoid the locals, with whom I'd embarrassed myself over the years in some of my more tragic attempts to conduct my daily life in their language. But I was trying to set an example for my children: that making mistakes was simply part of the process and nothing to be embarrassed about. Besides, as time wore on avoiding everyone with whom I'd made a language faux pas might have meant going without haircuts... and food...

Belgian cooking is a well-kept secret. My husband likes to refer to Belgian cuisine as 'French Gourmet meets German portions!' We enjoyed the opportunity to try Belgian foods at area restaurants and recipes when cooking at home with local ingredients. Wild game – venison, pheasant, wild boar, rabbit, etc. – are very popular during the autumn hunting season, and I decided to try cooking pheasant at home. This came about by a request from my children, who listened endlessly to a book on tape by Roald Dahl about a man who poached pheasants and raved about how delicious they were roasted. My girls wanted to try them.

With time, I'd learned from my mistakes and so before heading off to the local butcher, I looked up the word for "pheasant" in my faithful French dictionary. Of course, when I got there the word disappeared from my brain. I turned to my visiting mother-in-law (who is supposedly fluent in French), looking for the word. She supplied the word 'paon,' and that is what I asked for. The assistant behind the counter looked at me like I was slightly crazy and the conversation continued in French. She asked me to describe the bird I wanted and I proceeded to describe a bird with large, colourful feathers. She asked, "What are you going to do with it?" I replied that, of course, I was going to cook it and eat it. She shook her head as if to suggest that I was completely mad, and disappeared into the back to bring out the butcher himself. The same line of questioning was repeated. Finally, the butcher – in a flash of brilliance – remarked "Un faisan! Vous voulez un faisan!" ("A pheasant! You want a pheasant!") Back home once again, I appealed to my well-worn dictionary to find out what had gone wrong this time. There it was: 'paon'… "Peacock." Well – two more words to be permanently added to my French vocabulary bank. Ten points to my mother-in-law for helping me out on this one!

There are other, less minor language violations, which I admit to making over the three and a half years we lived in Belgium. But, along the way my French did improve by leaps and bounds and I was even able to pick up some 'localisms'… like the Belgians' catch-all use of the word 'normalemen' ("normally") to explain why a normal situation could not be, might not be or wasn't – normal. Essentially it absolved anyone of any responsibility when things went wrong. Normalement, I will be at your place tomorrow at 9:00 a.m… (But – if I'm late, maybe it's because I'm chasing a deer out of my yard)… Normalement, the washing machine should not make a noise like that (but… well, it's broken, because someone put a peacock in it!)… Normalement, your car should have been repaired by now (but, well… the part didn't arrive, so it's not ready)… etc, etc.

From "normalement" it was an easy step to adopt the distinctly Belgian "nanante" for "90" as opposed to the 'proper' French "quatres-vingt dix". This went over particularly well when I inadvertently used it in France one day and was met by a pitying stare, as there is little love lost between the Belgians and the French… and I had just identified myself as a Belgian 'peasant'. At times like this I revelled in the fact that at least someone felt my French was competent enough to mistake me for a Belgian peasant. Slowly my French "faux-pas" became less frequent and I was indeed able to communicate effectively in my day-to-day comings and goings.

My stumbles along the way to a state of semi-fluency provided my family with a number of stories and memories, which have remained with us long after our departure from Belgium. Although my kids were initially mortified and then keeled over in total hysterics at these mishaps, I tried to instill in them a sense of confidence and courage that enables the willingness to simply try, despite the potential for error. I also assured them that those same mistakes had brought me words and phrases that I would never forget and thereby strengthened my French. Next time we travel to Belgium on vacation, perhaps I'll let the kids order their own pheasant and see what we end up with for dinner! Mind your language… You never know what can happen!

To awaken alone
in a strange town
is one of
the pleasantest sensations
in the world.

Dame Freya Stark

Have Legs, Will Travel:
A Walking Tour In France

Heidi Lloyd-Price

From the first moment I learned of its existence, I was captivated by the holiday concept: a vast web of well-trodden footpaths, the freedom to start and stop where one fancied, a different charming village at the end of each day, great food and wine to be enjoyed with sharpened appetite, and the simplicity of travelling on one's own two legs. The Grande Randonnée (GR) is an extensive system of well-marked walking trails connecting one village with the next that meander all over France. I first heard about it in March 2004, when three friends and I stumbled upon a short, picturesque segment in the vicinity of Gordes, while attending a cooking course in Provence. That small hike was a tempting hors d'oeuvre, but I wondered whether my 100 lb. frame would be up to carrying a heavy pack for longer distances. When I learned that some GR tour companies ferry clients' luggage to the next destination by taxi, that clinched it – I was ready for the main entrée!

Back in Canada, I called my British fiancé in Germany to ask if he would join me on a summer GR adventure. We were newly engaged and had not yet made many trips à deux, except to visit one another on our different continents. Professor John (200 lb. frame, sedentary life style) was not immediately enthused about undertaking daily 20 kilometre treks, but was willing to give it a try. We were both 58 years old at the time, so it was reasonable to be slightly cautious. So, I embarked on what I thought would be an easy Internet search to find an appropriate tour. My research turned up plenty of personal reports about various sections of the GR (most were hopelessly outdated), but I had difficulty finding current information on tour companies. Time was running out and my hope was dimming to find something for our holiday. Fortunately, some Canadian friends had just returned from a GR trip and they provided key contact data. And just in the nick of time: even though it was already mid-July, I was able to phone 'Discover France' and arrange a tour in the Dordogne Valley for the first week of August.

The Dordogne Valley is located to the east of Bordeaux and is famous for its wines. The itinerary consisted of six days' walking (totalling about 120 kilometres) starting at Souillac and staying overnight at Cales, Rocamadour, Meyronne, Carannac, Loubrassac and finally St. Ceré.

The initial elation of having secured the booking was immediately followed by a pang of doubt as to whether our well-used legs would appreciate the last few kilometres of the daily 20 kilometre treks and/or indeed whether our temperaments would survive the rigours of facing the elements every day. After all, we weren't spring chickens and we hadn't undertaken anything that challenging together before. Ah well, nothing like a little adventure and suspense! I took out the 'silver' trip insurance policy just in case...

Most out-of-country clients would probably arrive by train from Paris, but John and I drove in his car from northern Germany, stopping to visit some friends on the way. Finally the day came and we carefully parked the car in the hotel garage at Souillac as had been pre-arranged. Most of John's reservations about the tour quietly evaporated on perusing the gourmet menu on the first night. The French do know their food and one can be served a spectacular meal in the most undistinguished setting in any small town or village. John's theory is that there is not only a tradition of valuing excellent food here, but also a tendency to noisily criticize any deviation from this expectation. Our plan was inclusive, but we always had the option to pay a little extra to upgrade if a particular dish or wine struck our fancy. Yes, things were looking promising and we clinked our wine glasses to the adventure about to unfold.

That first evening was very hot and humid. During the night there was a violent thunderstorm and in the morning it was still drizzling. 'How dare it rain on our first day!', I thought. Nature must have been listening, since, by the time we checked out after breakfast, the rain had stopped. With our lunches stowed in light backpacks, off we went. My diary says it was 9:30 a.m. when we left the hotel… and that at 9:31 a.m. we were already lost! Je m'excuse – temporarily disoriented. But, that was because each of us thought the other was following the map! After that we made it mutually clear who was doing the navigating. Soon we were back on track and walking along the Dordogne River.

Day one was steamy hot and when we came to a particularly inviting spot along the river, John suggested we take a refreshing dip. There were no handy changing rooms, of course, so the clothes were simply shed onto the rocks. Since there was a constant procession of canoes gliding by, each filled with tourists, we slipped into the water discreetly between boats, and then swam just like two rainbow trout against the current.

Refreshed, we started on a long uphill climb to Cales. But in the heat of the late afternoon, we were soon sweating again profusely. After what seemed like a long time, John reckoned we still had another two or three hours. However, just over the next rise we happily realized we were almost at our destination. We had survived our first day! That evening we indulged in a special bottle of wine, a Lussac St. Emilion 1975, with our dinner, which included the local specialty of sanglier (wild boar).

In general, our route stayed fairly close to the Dordogne River as we walked in an easterly direction, but included some loops to the south. The terrain was varied – sometimes we walked on small, tarred, rural roads, but we also followed farmers' field tracks and footpaths in the forest. I suspect some sections were remnants of old Roman roads. Although there were several steep climbs and descents, most of the way was reasonably gentle. The official rating was a level two – for occasional walkers. Each day we had a set of instructions to follow along with a map. A typical entry was: "Next crossroads turn left towards Carannac. Walk up this road. Turn right on the top 30 metres further, take a left and continue walking through a dirt track in front of you."

For those sections that followed the GR proper, there was the added help of the red and white trail markers. I think the challenge of finding our own way added a unique and exciting element to the holiday, keeping the brain somewhat engaged. It was like a treasure hunt, with a pleasant hotel and panaché waiting as the reward. I readily admit that two heads were better than one for map interpretation. We made a game of betting to see who could more closely predict our time of arrival at the next hotel; the loser bought the drinks.

All along the way, we were delighted by an abundance of butterflies – swallowtails,

blues, admirals, painted ladies, sulphurs and others. We were also serenaded by the buzzing cicadas, which are so famous in the south of France. Yet another bonus was the discovery of sweet, sun-ripened blackberries, which were thick on bushes framing the country roads. Try picking those delectable treats when speeding by at 120 clicks! All this we enjoyed to ourselves, meeting hardly any other tourists on the trail. Bliss!!

Day two was memorable for several reasons. It started out dry, but almost as soon as we resumed the trail, there came a few wet drops. This time we were not getting a pass on the rain. We were quite damp by midday when we reached LaCave, but decided to take the guided tour of La Grotte, nevertheless. Lascaux may be the most famous of France's caves, but there are many others in the region as well with sophisticated tourist facilities. The caves are marvelous for their stark physical beauty and also for the anthropological significance of their occasional occupants of long ago – Cro-Magnon man. Deep in the chilly La Grotte, we didn't see Cro-Magnon artistic displays, but we did witness something completely new to us – with the cave illumination temporarily switched off, the guide beamed UV light onto the slowly dripping stalactites –at each tip was an otherworldly glow. The effect was that suddenly we were enveloped in a subterranean starry night. The explanation involved the UV light being first captured by the calcite solution, and then being released at the point of crystallization. The guide poetically called it a beautiful demonstration of how minerals are 'alive.' We were in France, after all.

When we emerged at the surface again, the low clouds were producing a luke-warm, heavy drizzle. We never get rain quite that insistent in Calgary! As I recollect, my rain gear was not completely impermeable and John had forgotten to pack his. So, the main thing was to keep moving. In fact, that became our motto – 'keep moving!' By the time we reached Rocamadour just after 6:00 p.m., we resembled very much something the proverbial cat dragged in. Tired, hungry, and sopping wet, we could imagine how the pilgrims in the Middle Ages had felt arriving at this holy of holies in the centre of a church and chapel complex carved high into the solid cliff face.

It is a very nice feeling to walk into the destination hotel lobby and not have to worry about parking. It is even nicer to be shown to your room and find your bags already neatly installed. And, it is really, really nice when you arrive feeling so tired and sodden and discover your room is equipped with the world's most luxurious bathroom. I doubt that many of the previous occupants could have appreciated the Jacuzzi with gold-plated fixtures and super hair dryer more than we did that evening. Memorable, indeed.

Thank goodness the prediction of continued torrential rains never materialized, and for five more days we enjoyed good summer weather. Our legs responded well to the challenging regimen – it's remarkable how quickly the body can adapt. Needless to say, we had good appetites by the end of each day, so the excellent cuisine offered by our hosts could be suitably relished. Some of the foods we enjoyed were escargots, pâté de foie gras, confit de canard, magret de canard, calves' liver, stuffed rabbit and fresh fish such as Lotte, Sandre, salmon and trout. This was all washed down with some excellent local wine. And, how romantic it was, too, being able to dine outside in the long summer evening light.

We knew we were in wine, duck and cheese country, but the revelation for me (more than for European-born John) was that we were also in nut country. Our itinerary took us along the route de noix and we must have seen a thousand walnut trees! Indeed, the local specialty was vin de noix, which was served as an aperitif.

In Canada, we enjoy some of the most spectacular natural scenery in the world, and the wild, rugged Rocky Mountains are literally in my home backyard. But the Dordogne experience offered something different and invigorating. As well as being an exciting trip to another country, our self-guided walking tour was also an inward journey for John and me, and we seemed to have made all the right turns there, too. More regions of France remain for us to explore on the GR and, when that is finished, I've heard about a wonderful trek on a mountain called Kilimanjaro…

Got Spice?
Friends Together for a Cooking Week in Provence

Heidi Lloyd–Price, Beth Powell, Cynthia Russell & Stefa Williams

Last fall, the e-mail said the three were booked for a week's cooking course in Provence in March and would I join them? My three ex-roommates from Montreal McGill days, that is. Since graduating about 30 years ago, we had pursued different paths, leading to various careers, husbands, families and distant homes. Beth and Stefa were in Toronto, Cindy, in Vermont and me, Heidi, in Calgary. Now we were in our fifties (how did that happen so fast?), the children were grown and it was time to re-connect.

Well, how could I not go? My intuition said this was an opportunity not to be missed. The funds were there, fortunately, so I signed up for the week at Les Mégalithes, which is just to the east of Avignon, a region made famous by Peter Mayle in his colourful book '*A Year in Provence*'. Cooking is not my top priority, but I figured a few classes wouldn't kill me and it would be fun to go to France to see my long-lost women friends. Beth is a doctor at a large Toronto hospital (three boys), Stefa, a retired chartered accountant (two girls and a boy), Cindy, a teacher and education grant administrator (a boy and girl plus three more children from a second marriage), and me, an erstwhile cell biologist, but now full time household manager after being widowed four years ago (two boys).

Just the thought of the upcoming trip brightened the short, dark days of winter. As departure time approached, there was a flurry of e-mails about TGV schedules and aprons. Efforts were made to upgrade my French by reading Pagnol and watching videos. At the last moment, there were fleeting thoughts of – Would we have changed? Would we argue about what to see and do? Would I burn the food?

Finally, there I was in the Avignon train station. Cindy was the first to arrive via Nice, and then, soon after, Beth and Stefa from Paris. We were overjoyed to be re-united! But, we couldn't get too relaxed yet, because we still had a destination to make. The immediate challenge was to find our rental car and fit our combined luggage into it. I am still amazed how we managed this – with even a few cubic centimetres to spare! This accomplished, we set forth into Avignon, smugly equipped with a computer printout of map quest directions for escaping the city.

Unfortunately, the initial road signs we encountered didn't seem to correspond very closely to those instructions and we had to resort to rusty French and sun-sightings to get oriented. Soon we were driving along the desired route, breathing a sign of relief and already drinking in the beauty of the early-springtime countryside. In mid-March, the green leaves were not yet out, being the cherry and almond trees were in blossom and delighted our North American winter-weary eyes.

Les Mégalithes was even more picturesque and charming than advertised. Sheltered midst old pine trees on an outcrop overlooking vineyards and olive groves, the well-maintained stone farmhouse with swimming pool (still canvas covered in off-season) was the essence of Provence. Sylvie Lallemand, our host and teacher, was waiting at the door to greet us. After we drew lots as to who got which bedroom, Sylvie did not waste time in putting us to work making bread for breakfast. Both Stefa and Cindy had brought aprons for each of us and one set of these was immediately commissioned.

During the week we cooked fish, lamb, beef, chicken and rabbit dishes and prepared accompanying vegetables as well as hors d'oeuvres and desserts. All the chopping and mincing required a certain level of wine consumption and we soon acquired a taste for the local Côtes du Luberon and Côtes du Ventoux vintages. Thus lubricated, we learned not only about how to knead bread dough (not too much!) and to poach fluffy egg whites (not too much either!), but we also caught up on each of our life's stories and shared advice about boyfriends, lovers, husbands, ex-wives, chick-flicks, what was really going on in the recent set-in-Provence movie '*Swimming Pool*' and shoes. Mais qui, never underestimate the importance of a pair of elegant black shoes! We soon discovered that Sylvie, who was happy to practice her English, could hold her own on the non-food topics as well. Each evening there was another layer of history revealed and we began to wonder how old Sylvie could be to account for the richness of her experiences.

Everyday the four of us went out on excursions to see neighboring points of interest. Where beauty and history permeate the very air you breathe, it was a challenge to select just a few of the many possible places to visit. Also, we discovered that, in addition to its wine production, this region of Provence is particularly noted for its superb olive oil and lavender. On our second evening we held a palaver to draw up a list of sights to see and in what order. Fortunately, we had no problem in reaching a mutual agreement on a tentative plan.

Travelling in the car was made easier by the fact that we had self-organized our respective roles from the start. Thus, Beth was our designated driver, since she had signed for the rental car and Heidi was the map-holder and navigator (any wrong turns were due to poor signage, of course!). Stefa and Cindy had travelled previously in Europe with Beth and thought it prudent to nab the back seat! Cindy leaped out of the car whenever we needed to park and guided us into or out of the exceedingly narrow slots that pass for parking spots in Europe. Stefa's job was to restrain herself from making acerbic comments about the performance of any of the foregoing endeavours and was thus probably the most difficult task of all.

We had one interesting near-incident where we almost left a small sample of fender paint on the stone wall comprising one of those impossibly tight corners that typify the heart of French towns. It was our first day and we had spent the morning and early afternoon in the market of L'Isle sur Sorgue. One slightly wrong turn (another instance of poor signage, of course) had us driving into the now-vacated market instead of away from it and, to our growing concern, there was no way to turn around. The ever-narrowing maze of cobblestone streets delivered us ultimately to what appeared to be a prohibitively tight corner. And... now there was a truck behind us!

After some minutes of assessment and subsequent gesticulations on the part of our nimble parking authority Cindy and tentative maneuverings by Beth (whose thought bubble was easily read to include the possibility of many extra euros on the car rental invoice), the female passenger from the not-so-patient truck suddenly appeared next to us and exclaimed in very fast and exasperated French something which in general was not very complimentary about the road skills of Canadian drivers. This helped our corner

Elizabeth Powell, Heidi Lloyd – Price, Stefa Williams & Cynthia Russell

reconnoitering tremendously, of course.

A few minutes later, with tremendous reserve and aplomb to her credit, Beth slithered the car around the corner with not even a single scratch as a souvenir. After we regained the proper road and got over our indignation of the verbal assault, we wondered how we were so readily identified, as Canadians and what were our compatriots doing on French roads to deserve such a reputation anyway?? Never mind, so kindly had other French people treated us during our stay, that we could forgive one grumpy citizen with a bad-hair day.

Since we didn't want to miss our afternoon cooking class, time constrained the amount of sightseeing we could do. Nevertheless, we managed to visit the Locoest chateau (a sometime residence of the Marquis de Sade), the picturesque towns of Bonnieux and Roussillon, and the restored stone huts of Les Bories, Saint Rémy with its famous plane tree tunnel, Les Baux and the massive Palais des Papes in Avignon.

We had two trips to the nearby medieval hilltop village of Gordes because we liked it so much. The second trip was by foot; following a hiking trail that started from the top of the hill behind Les Mégalithes. What a treat it was to wander along this well marked path that took us over the mountain and then down through the vineyards and olive groves below. The spring sun was actually quite hot and we were even down to T-shirts at times. Sylvie recommended her friend's restaurant in Gordes and we were able to sit outside in its lovely café terrace garden for lunch.

We returned from our adventures in late afternoon (later and later on succeeding days, as it turned out) to begin the dinner preparations. Sylvie would be waiting for us with all the ingredients carefully laid out. She was very well organized having taught cooking at Les Mégalithes for over twenty years! At the end of our stay, Sylvie presented us with champagne and diplomas. We all had passed! Of the many dishes we prepared I think the Fresh Asparagus with Hard-Boiled Egg Dressing was outstanding. And our last dinner, an Aioli with codfish, really surprised me, as I didn't think a cold dish (fish with assorted pre-cooked veggies) with garlic mayonnaise could ever be so delicious.

All our recipes can be seen in Sylvie's book called '*The Tastes and Colours of Provence*'. This gem also includes a description of some of the places we visited in the area and is illustrated with Madame Lallmand's own watercolours, the originals of which we saw hanging in her house. Sylvie was happy to autograph the multiple copies of the book that we ordered. Since 9/11, she said, her business, along with many other tourist dependent enterprises in the region, has suffered, as North Americans have become more reticent to travel.

Had we women friends changed from our university days? Not really. I think our temperaments and idiosyncrasies are still pretty much the same, perhaps even a little amplified over time. (Only in a charming sort of way, of course.) We got along just fine, just as we had so many years ago. Although most of our tears were from laughter, some of our stories were poignant, such as the recent stroke suffered by Beth's husband. He had bravely encouraged her to go on this trip anyway, and it gave her much needed rest and support. We were all reminded that life and health must never be taken for granted.

There are many cooking schools in Europe and I'm not sure how we chanced upon Les Mégalithes, but it was an unqualified success. Madame Lallemand, as our knowledgeable teacher, gracious host, raconteur and ageless model of French femininity, was an inspiration. We also appreciated how economic Sylvie's wine cellar was.

Going in March definitely had its advantages. The fruit trees were in blossom but the tourist crowds were not. The shopping urge was amply satisfied in the village markets where colourful Provençal textiles were partout. Among our collective purchases were tablecloths, earthenware pottery, a tapestry jacket, silk scarf and a dozen or so bottles of Châteauneuf-du-Pape. We also stocked up on ceramic garlic-grating dishes, attractively glazed in sunflower colours. Much to our dismay, Sylvie didn't approve of them (you lose a lot of garlic on the rough grating center), but never mind, they make great olive severs! Being a notorious bug-lover, I also couldn't resist a couple of those ceramic 'cigales', which are quintessential Provence.

Which reminds me, I'll have to return to Provence in the summer to hear the cicadas sing. In the meantime, the photos, journal, a well-used white apron embroidered in yellow thread with '*Les Mégalithes 2004*' and the occasional cooking session will serve to refresh fond memories of our week in Provence.

Heidi Lloyd-Price is from Calgary, Alberta, Beth Powell is from Toronto, Ontario, Cynthia Russell is from Cabot, Vermont and Stefa Williams is from Oakville, Ontario.

Where Galileo stood at night to take
The vision of the stars, we found it hard
Gazing upon the earth & heaven, to make
A choice of beauty

From 'Tuscan Bellosguardo' by Elizabeth Barrett Browning

Photo above is the Italian coastline.

Saudi Arabia's Wardrobe Essential: The Black Abaya

Linda Barghoorn

A woman's wardrobe in Canada usually contains several essential pieces such as a simple black evening dress, a flattering bathing suit for summer days at the cottage or poolside, a comfortable pair of jeans for everyday… It is often items such as these that we feel we could not function without – and they take into consideration things like comfort, personal style and current fashion trends. When we transferred from Toronto to Saudi Arabia due to a career move with my husband's job, I traded these in (or more appropriately, I covered them up) for what is the Saudi Arabian woman's wardrobe essential: the black abaya.

Our first visit to Saudi Arabia was a whirlwind trip – a few days in and out – to check out housing, schooling and to get a feel for what our new life was to involve in a country that is often shrouded in mystery. For this occasion I was provided with a "loaner" abaya upon our arrival at the airport in Saudi, from a woman who was assigned to show us around. The garment I was given was a simple affair: a full-length black robe – much like a graduation gown – which zipped up the front and covered me entirely from neck to wrist to toe. Although I'm sure one of my first impressions must have been of the somber crowds of women sheathed similarly (and often more severely) in black wherever we went, I remained somewhat oblivious at this stage – caught up in the excitement, anxiety and curiosity of what my life was to become. I'm also pretty sure it was stifling under there in the 40+Celsius heat; but the memory of that, too, is vague. I was simply too overwhelmed with everything else I was taking in to take much notice!

Having accepted the offer, we returned several months later to our new home in Jeddah, on Saudi Arabia's Red Sea coast. I was still armed with the simple, black loaner that was to be my companion for the first few weeks. As our plane entered Saudi Arabian airspace, a wave of movement swept through the plane. A swishing of fabric rustled the air as the women around me pulled on long, black folds of cloth and covered their hair with matching black scarves. What had only minutes before been stylishly dressed females became shapeless human bodies cloaked in black. Taking my cue, I pulled on my abaya, as my young children sat watching my transformation in fascination. Luckily, they were still too young to have to adopt this practice. I left my headscarf off and sat back down, in my black robe, but with my brown hair and blue eyes still proclaiming loudly that I was a foreigner in this place.

The reality of what I had gotten myself into began to dawn for the first time. I shot my husband a quick glance of mixed humility and hopeful defiance as I assumed my new profile. More than anyone in our family, I had been eager for this opportunity to experience a new culture, leaving behind family, friends and a blossoming career. I was prepared to be open-minded, but parts of me wrestled with this black robe that, for many of us with western eyes, is a symbol of women's submission.

We stepped off the plane into the blistering heat of an August day in Jeddah, Saudi Arabia's second largest city. Drops of sweat began to slither – slowly, annoyingly – down my back. The abaya wrapped itself stubbornly around my ankles, as I tried to make my way down the metal stairway to the shimmering tarmac where the shuttle buses waited to take us to the airport terminal. Gathering up the extra folds and holding them up enough that I could keep from tumbling down the stairs, I proceeded to shuffle my way onto the bus and took a deep breath. This garment seemed to be taking on a personality of its own - morphing into an awkward burden, both physical and psychological. The air conditioning in the airport was a welcome relief to the steady trickle of sweat down my back. Settling into the privacy of our hotel room an hour or so later, I have to admit to throwing the abaya onto a nearby chair with great relief… and frustration!

"O Prophet, tell your wives and daughters and the women of the believers to draw their cloaks all over their bodies, that is most convenient that they should be known (as such) and not molested," says the Koran. It directs women to cover in order to protect themselves within society. The practice of covering the head and body, known as hijab, originates from the Arabic word 'hajaba', which means 'to shield' or 'hide from view.' This includes women's hair, which – much like exposed female breasts in western society – is considered inappropriate in public.

The Saudi Arabian form of hijab is among the most extreme in the Muslim world, requiring women to cover from the neck to the wrists and ankles, in a black abaya, with a headscarf covering their hair. The garment should be loose enough not to lend itself to the shape of the woman's body and thick enough not to show the colour of her skin. From here local customs and individual beliefs dictate whether the veiling goes still further. For those who follow the strictest version of hijab, veils, gloves and thick stockings cover faces, hands and ankles so that not an ounce of skin is revealed.

Supporters of the hijab argue that it allows humans to treat each other as equals without being distracted by superficial beauty. To a western eye, it seems to eliminate that very humanness, on which we place such high value. For the devout Muslim woman, the practice of veiling is a sign of her modesty and chastity, and elevates her to a higher level than the woman who allows herself to be admired for her beauty and sexuality. But for many of us from the west, the veil conjures up images of subservience and inferiority, in a country where women are forbidden to drive, work alongside men, or even socialize in mixed company. Hijab must be observed whenever in the company of a male who is not an immediate family member – that is: a father, a husband, or a brother. For most expatriates living in Saudi Arabia, the abaya was simply a ticket to move around the city without being starred at, harassed or detained by the mutawwa, the Saudi religious police. While it would never play a role in my religious beliefs, I could accept it as a means of conforming to local standards. As they say: "When in Rome…"

We spent several days in the hotel before being able to move into our home. While at the hotel, every excursion from the room required the black abaya – to make an inquiry at the reception desk, to have a meal in the restaurant, and to go for a walk with the children. My young girls caught on quickly and were vigilant to remind me that I mustn't forget it in the room; horrified their mother might break the dress code and get into trouble with the authorities! Soon after, we settled into our new home on a large expatriate compound, shielded from the surrounding neighborhood by a wall two-metres high. Inside those walls we were free to walk around as one would in the west, exposing the white expanses of our arms and legs in hopes of getting a tan and free of any need for the black abaya. It was tempting to nestle inside this cocoon and avoid any excursions out into the 'real' Saudi Arabia. And for a short time, that's exactly what I did – fussy over unpacking, setting up our home, enjoying the beautiful blue swimming pool just outside our back door and meeting our neighbours. But life in a Club-Med type existence can become tedious rather quickly and I knew that I hadn't moved here to live in a cocoon. Besides, as I was expected to return the loaned abaya that had been provided for my arrival in Jeddah, it was time to venture out into the city and purchase one of my own.

And so, on a Thursday afternoon - the beginning of the Saudi weekend - my husband and I packed our family into the car and headed downtown. Along the way we passed a small outdoor mall, where a glass-fronted shop sported an impressive collection of abayas. While my family waited inside the jeep, I hopped out quickly to explore what was on offer. My first lesson came quickly: although all abayas came in the basic monochrome black variety, some were clearly designed – and priced – for those with expensive habits (and the wallets to match!). With little to differentiate them except some beadwork and embroidery, I could not understand what merited price tags that started in the hundreds and marched into the thousands of dollars. I was looking for something in the $19.95 range (as that was the approximate guideline I'd been given), and was clearly in the wrong place.

I browsed casually for a few minutes, feigning interest and then quickly exited the store. As I climbed into the car I explained that this must be a "designer abaya" store – something I hadn't known existed until that moment. My husband was no more in the mood to spend a small fortune on what he called 'the black robe' than I was. And so we headed for the souk, or marketplace, in the historic quarters of Old Jeddah.

The modern highway leading from the newer suburbs into the old city gave way to a myriad of narrow, twisting alleys, marking the beginning of the souk. Most expats preferred to shop in the modern, western malls in the more affluent areas of Jeddah. Here, in the souk where the locals did their shopping, there were few western faces to be seen. Faceless, black-sheathed women made their way through the tiny pedestrian alleyways lined with shops crammed haphazardly with goods from floor to ceiling. Determined to be the independent western woman I am, I stepped out of the car and arranged to meet my family again in an hour at the car park. By then I planned to have my own new abaya in hand.

The souk was just beginning to open following midday prayer, and the rattle of rusty, metal shutters being pulled open clattered down the alleys. A cluster of abaya shops lay ahead. Preparing for a lesson in humility (and a test of my basic Arabic), I chose a shop and stepped inside, greeting the shopkeeper with a muted "as-salaam aleikum." The long robes hung simply from racks and hooks on the walls thick and black, like enormous shrouds. No mannequins or fancy merchandising here. These were the abayas for the masses, with few hints of elegance or style, and no fancy price tags.

The owner selected several, which I indicated by pointing at various hooks high up on the wall. In the end, I selected something simple with a bit of black embroidery, which closed from neck to toe with a series of practical metal snaps. We discussed the price and I negotiated the obligatory Middle Eastern discount with a brief exchange of basic Arabic. Ecstatic at my accomplishment and the small sense of independence gained in completing this adventure on my own, I made my way back to the car.

As the first weeks passed, I began to adapt to the oppressive heat under the thick, black material. At the same time the abaya offered me anonymity and freedom of movement. It became an automatic part of my wardrobe when leaving the compound – always hanging by the door ready to be snatched up in a hurry when walking out the door. I began to resent it less, and accepted it as a ticket to move in public without attracting the attention of the local population. While my uncovered hair still alerted that I was a foreigner, I was relatively successful otherwise at blending into the crowds. I always carried the tarha, or headscarf, with me in case of a mutawwa sighting, but could not bring myself to completely cover. For me it represented a line, which I could not bring myself to cross, as if completely covering somehow swallowed my entire person.

The mutawwa, formally entitled the 'Committee for the Promotion of Virtue and Prevention of Vice', is a rather undefined organization of strict Muslim men. Although its title sounds Orwellian, it does indeed exist. It is designed to ensure that the strict code of conduct outlined by the Koran is followed in society: that the faithful go to mosque to pray at the designated times, that the strict segregation of men and women is enforced, that materials (from books to DVD's to Barbie dolls) are appropriately censored, and so forth. An encounter with the mutawwa, although they are technically "unofficial" in their role as religious police, can nevertheless be notoriously unpleasant, and was therefore something to be avoided at all cost.

Increasingly I spent more time outside the compound and my faithful black abaya, was always with me – on photography excursions into the old city quarters, on weekend camping trips into the desert with friends, and formal evenings at the consulate in town. That abaya covered up everything from bathing suits, shorts and halter – tops to fancy evening dresses; but eventually it started to look rather tired and tattered around the edges – like a child's favourite teddy bear, which has seen better days. This, at least, was evidence of how much my abaya was being used; that I wasn't cocooned in our compound, but rather had taken the opportunities to get out into society – to explore, to learn and to see for myself. I grew to appreciate at least some of the aspects of this culture and society and found it to be a fascinating place.

A year after our arrival in Jeddah I announced that I was going shopping for another abaya. Although it seemed a normal event to me, my husband seemed rather taken aback that I felt the need to expand my collection. But in addition to the fact that my abaya was slowly disintegrating, I had become aware of the fashion changes in the abaya world... or maybe mine had always been out of fashion, and I'd just been hopelessly naïve in the first place. Whatever the case, I now understood that as the only piece of clothing, which I presented to the outside world, the abaya spoke volumes about the person underneath. Mine said things like "peasant," "pauper," and – probably – "totally and hopelessly unfashionable foreigner"! I realized that the abaya was more than a black sheath, which allowed me to go out in public; that it revealed as much about my sense of self and individuality as do regular clothes in the western world. There were cheap, functional abayas that looked just that. And there were more expensive, elegant ones, which afforded the only hints of individuality and style, which the Saudi public was allowed to see. I knew which image I wanted to portray, and "peasant" wasn't it!

This time I did not return to the old souk downtown for my abaya. I chose a stylish – but not designer – shop in a local mall near home and took the time to assess styles and sizes. Although the overwhelming majority still came in basic black, several new colours had snuck into the palette: midnight blue, grey, burgundy red (which seemed rather racy in our strict surroundings). I chose one with metres of extra fabric in its flowing, graceful lines. It was decorated with delicate black embroidery and beadwork, and closed with a single hook at chest level, which allowed it to drift open rather alarmingly. This necessitated that I take hold of the extra fabric and carefully hold the front closed when I walked to avoid revealing any signs of my long legs underneath. And so now I had an "everyday" abaya and an "evening" abaya... although the dated, "everyday" abaya came out less and less. My Saudi wardrobe essential had just seen its first upgrade!

Our last evening in Jeddah - three and a half years after our arrival – brought a fitting, and fortunately relatively minor encounter with the mutawwa. After shopping at a mall, my husband and I walked through the parking lot towards our car. Having so rarely needed it, I had become careless with my headscarf and left it at home. The individuals patrolling the parking lot had a glorious view of my short, brown hair blowing in the wind. Their large Suburban jeep began its slow approach. My husband hustled me off to our car and waited for the mutawwa. He listened politely to their lecture on the need for humility by "our women," before politely explaining our wish for a peaceful last evening in their country. Perhaps understanding that with tomorrow there would be one less rebellious foreign woman on their soil, they let us take our leave. We no longer live in Saudi Arabia. As we left the country for the last time, I sat on the plane wrapped in my abaya, which had accompanied me for three and a half years like a faithful companion. Surprisingly – unlike that first day in the hotel – I didn't feel the urgent need to strip it off and toss it aside. It's packed away neatly in a box now – like a preserved wedding dress – a reminder of a significant and fascinating time in my life. Occasionally, I pull it out to remind me of its role in my initiation into a culture that seems so severe by western standards. Despite the restrictions of living in Saudi Arabia, it was a wonderful opportunity that allowed me to learn and experience so much on a personal level. I left a corporate career behind, and gained several part-time careers, which I've carried with me in the years beyond Saudi Arabia. I was also fortunate to meet some wonderful Saudi women and had an opportunity to see what their life was like behind the veil.

I've traded the Saudi Arabian wardrobe "must-have" back in for the comfort of my western wardrobe essentials: my favorite blue jeans and a comfy warm sweater. We are now based in London England – one of the big cities of the western fashion world. I don't have to look over my shoulder anymore for the religious police we anxiously avoided in Saudi. Now it's a different kind of fashion police I have to be on the lookout for. But, fortunately, in this fashion capital, almost anything goes!

Too often…would men boast
only of the miles covered that day,
rarely of what they had seen.

Louis L'Amour

The Kamikaze Taxi Drivers of Abu Dhabi

James Mac Laughlin

Abu Dhabi

I'm a college English instructor who works in Abu Dhabi, the capital of the United Arab Emirates. Abu Dhabi is the lesser-known sister of Dubai, which is two hours away by car, up the gulf coast. Despite its lack of notoriety, Abu Dhabi remains the center of wealth and power in the Emirates owing to its huge oil reserves.

The population of Abu Dhabi is about 600,000. I don't know how many cars are on the road. My guess is about 150,000. The city itself is quite modern. However there is quite a disparity between the haves and the have-nots. The have-nots such as the taxi drivers I am describing are guest workers who actually make up the bulk of the population here. The city is undergoing quite a bit of development at the moment though not on the scale of Dubai.

People often ask me if I feel anxious living in the United Arab Emirates. In fact, many people react with alarm when I say the country's name. I suppose the word Arab conjures up images of masked gunmen and roadside kidnappings. In fact, the country has a remarkably peaceful history, practically free of political violence. That does not mean, however, that there is no danger.

Every day, I take a serious risk getting to and from work. Taxi drivers here (as with many drivers in the UAE) have a completely alien notion of acceptable risk. Speed itself seems to be the goal rather than saving time getting to a destination. Speed limits are in place but no one pays any attention to them. Speeds of 120 kilometres per hour are not unusual on city streets. Taxi drivers constantly tailgate and flash their lights at slower cars. (Taxis are usually Toyota Corollas.) Upon passing, a taxi driver will gesture to a slower driver as if to say, "Are you crazy or something?" When approaching a red light a driver will speed-up rather than slow down. When the signal changes, a typical driver leans on his horn – telling the driver in front that he is starting too slowly.

The inherent danger of this type of driving is increased by the absence of seatbelts in rear seats. I usually sit in the front where I can strap on a seat belt, get a good view of the action and yell out warnings to the driver. However, on occasion, a driver will have a terrible case of BO making me wish I were in the back.

The positive side to all this is the price. You can go anywhere in the city for under five dollars Canadian. In fact, my seven-kilometre ride to work costs me just over two dollars, the cost of a bus ride in Vancouver.

James Mac Laughlin is originally from Vancouver Island.

Shopping 'Middle-East Style'

By Linda Barghoorn

Shopping… it is a passion for some, an obsession for others, and for many it is a simple necessity; whether it is for groceries, a gift or a new pair of shoes. Some of us enjoy it, while others find it mere drudgery. Often, for many of us in the west, it can be a simple, rather impersonal event. Walk into a store, look for the appropriate item, (hopefully) find it, take it to the cashier, pay for it, and leave. In the traditional souqs, or marketplaces, of the Middle East, shopping is an entirely different ball game. When was the last time you went shopping and had a shopkeeper call out to you to come and sample his wares, offer you a cup of tea while you consider your purchases, and then argue with you over the selling price?

Shopping in a middle-eastern souq is like shopping nowhere else in the world. It can be simultaneously a stroll through history, a lesson in the skill of price negotiation, and an assault on the senses. And 2,000 year-old Sana'a, in the heart of ancient Yemen, is among the most magical of souqs! Isolated both politically and geographically until recent decades, the country has often been referred to as "the land that time forgot". A stroll through Sanaa's historic centre offers a fascinating glimpse into daily Yemeni life. It is also an "adventure shopper's" dream! Yemen was the home of the renowned Queen of Sheba and the crossroads for the ancient spice caravans travelling between Asia and Europe. Jewish silversmiths once lived and worked in Sanaa's alleys, creating beautifully worked pieces – lamps, tableware, jewellry. The rich, dark coffee known as 'mokha' originated from the mountain terraces to the south and became known the world over. And, so, many of the goods you will find here are unique to Yemen and things to be treasured back home!

For a westerner accustomed to the orderliness and sterility of our modern, North American carbon-copy shopping malls the first few moments in an Arabic souq can be shocking! Shouts of shopkeepers calling you into their shops compete with the clanging of metal in a nearby workshop, the laughter of children playing with a ragged ball in the next alleyway, and the occasional clip-clop of donkeys' hooves. Tempting aromas of exotic spices, rich, dark coffee and spicy grilled meats compete with sometimes not so tempting aromas… rotting vegetables left in an alley corner, uncollected garbage, and diesel fumes from a nearby generator. Brightly coloured bolts of fabric in the clothing alley and neatly stacked earthy-coloured pyramids of spices at the spice market tempt the eye. The alleyways bustle with activity and personal space is an unfamiliar concept, as shoppers in the busy alleys compete for elbowroom.

A majestic 16th century gate, known as Bab al-Yemen (the 'Gate of Yemen'), marks the entrance to the main square in the heart of the old city. Many of the buildings are built in the traditional Yemeni style: multi-storied homes built of sand-coloured stone. They resemble elaborate sandcastles, with intricate patterns in brick and paint decorating their facades and a line of jagged pyramids, like a row of shark's teeth, marching along their horizontal rooftops. In the night, with their illuminated stain glass windows, they resemble a skyline from the pages of '*A Thousand and One Arabian Nights*'. From the main square a multitude of narrow, twisting alleys zigzag away into a confusing labyrinth that threaten to swallow up the 'directionally challenged'. Tightly clustered miniature shops and stalls line the streets, their wares heaped and piled and overflowing – no fancy displays or merchandising techniques here!

Making our way down one of the alleys off the main square, we met Ali. Tall and slender, with pools of brown for eyes, he leaned casually outside the shop front where we stopped to make our first purchase. We struck up casual conversation, with Ali posing the typical local questions: "Where are you from?" "Do you like Yemen?" in his accented, but competent, English. When asked for directions to a shop we wanted to visit, Ali was quick to respond, "I will take you." And so, we headed off, weaving our way through the streets, with charming Ali leading the charge.

Before this begins to sound like the opening lines of a steamy, foreign romance novel, let me explain that Ali was 11 years old. Despite his years he had a wit and charm well beyond his years, and was clearly in his element, chatting and pointing out shops as we threaded our way behind him through the complicated maze of streets. It was quickly obvious that he knew the intricate system of alleys and laneways like the back of his hand; but he also claimed to know where to find the best leather, silver, spices, pashmina shawls, local coffee, antiques and any other of our hearts' desires - our own personal shopping assistant! It was clear that this was Ali's backyard and that leading tourists through this market, was something that Ali had done more than once!

And so we followed him... He led us from one shop to the next, discreetly waiting outside while we made our purchases, but not so invisibly that the shopkeepers weren't perfectly clear about how we'd come to find them. Ali would be returning later to collect his commission from each of them for the extra business he'd solicited that evening.

The souq is divided into sections selling similar goods, on the same principle as a modern department store, but resembling no such thing! Many shops are little more than an indentation in an alley, with stacks of merchandise piled high in the street. It does make it easy to find things, however, as all the fabric shops are in one area, all the house ware items in another, the antiques somewhere else. All the spice shops were clustered together, with deliciously exotic scents wafting from immaculately stacked pyramids of herbs and ground spices – turmeric, cumin, zaatar, saffron, among many others – in shades of brown, beige, yellow, orange, green and black. Beside them were chunks of frankincense and myrrh, which are burned like incense, reminders of the almost forgotten days of ancient kings and queens.

As we wandered, shopkeepers called out pleadingly to us to come and see what they had to offer. It is easy to be lured into the shops by their offers of "just look; no need to buy." But, they are skilled salesmen and negotiators. Once you are inside it can be very difficult to get out without having added another purchase to your collection. Particularly after you've enjoyed a steaming, cup of deliciously, sweet mint tea, which is the customary offering both in shops and in private homes. It is impolite to decline, and while it may serve as an effective ploy to keep you in their shop, it is also a wonderful way to engage in conversation with the locals. Shopping is a much more personal, intimate experience here than in the West. No simple "thank you, I'll take this," pay and leave approaches here!

Sanaa, the old city capital of Yemen. A UNESCO world cultural heritage site.

As we sipped our tea in the shop offering Indian pashminas at a steal, the shopkeeper pulled out shawl after shawl, in a dizzying array of colours and patterns, until our heads were dizzy with the selection. As we weeded out the ones we wanted from those we did not, he continued as he began to get a sense for what we liked: "You like this colour? I have in different patterns… You like this? Or this? You see? You like this colour instead? I have other colours. Maybe you like this one?" And on it went. It's easy to see how shopping can become a full-time occupation in a place where shopkeepers are eager to keep you in their shop for as long as possible, rather than eager to see you out the door!

But beware! While the treasures here may tempt you, shopping in Yemen is not for the meek and inhibited! Any price marked on an item is merely an 'indication' rather than a firm price and it is customary – no, it is expected – that you will negotiate the final price with the shopkeeper. Not to do so seems to rob the shopkeeper of a further opportunity for conversation with his customer, possibly over another cup of tea. But to be a successful participant you need to understand two things: what you are willing to pay; and that most of the shopkeeper's protests about how unfair your suggestion is, are good-natured banter and should not dissuade you from sticking to your guns.

The Yemenis are seasoned negotiators and proved to be almost unconquerable adversaries in some shops. The 'pashmina man' offered lovely shawls at a price that was hard to argue by Western standards. And so, although we did go through the customary negotiation process, we did not have to bargain hard for a price we felt was fair. The antique shops proved to be much more difficult.

At the Silver Souq, glistening shops offered up an Aladdin's cave of shiny treasures – vases, lamps, bowls, jewellry, candlesticks - piled precariously on top of one another in heaps and mounds. A thriving Jewish community in towns once produced antique silver jewellry across Yemen. Now these communities are almost extinct, making it increasingly difficult to find original, well-crafted pieces. Traditional djambias, the ceremonial daggers, are also very popular – both with the tourists and the local men, who tuck them into finely embroidered belts around their waist. This was the ultimate destination, and where we spent much of time, over cup after cup of tea.

I sifted through the treasures, made my selections and placed my desired purchases – a djambia and a few pieces of antique silver jewellry – on the counter to await the shopkeeper's sum. He presented me with an outrageous price, to which I countered with a proposal of one third that amount. He feigned shock and dismay that my suggestion would ultimately have his family starve from lack of livelihood if he were allowed to sell his goods at such ridiculously low prices. The negotiations began – an elaborate tango of exclamations, pleas, frowns, smiles, and frantic gestures. Always remember that – most of the time – the shopkeeper's shock and outrage at your proposal, is simply part of a good-natured banter between both parties. He's not expecting you to pay the full price, nor are you expecting to get away with one third. In the end, you'll meet somewhere in the middle, on a price that should make both buyer and seller happy. How much in the middle depends on who has the stronger will or negotiation techniques. He is not going to allow his family to starve by selling to you at a loss, and you are not going to buy something at a king's ransom.

So why do it? – insist many westerners who are frustrated by what seems to be an interminable process, which ultimately meets somewhere in the middle anyway? Because that's how it's done here! It's part of the experience of shopping in the Middle East, and a wonderful opportunity to get to know the local culture and the people. All the bickering, pleading and bantering along the way is just part of the game! It can be fun if you do it with a smile and see where your negotiation skills take you. It can even be addictive… When was the last time you had a long conversation over a cup of tea with the sales clerk at the jewellry shop in the mall in Calgary or Toronto or Montreal?

Darkness had fallen silently as we made our way through the souq, and lights and candles had slowly flickered to life to illuminate the stalls and shops. Suddenly we were plunged into complete darkness as the electricity was cut. Ali explained that this was normal, although I couldn't quite determine whether it was a conscious effort to conserve energy or simply an accident that took place with regularity. Sana'a, it seems, doesn't have enough electricity to supply its ever-increasing needs. Shopkeepers lit candles and gas lanterns at some stalls, while at others the low steady hum of small generators sprang into to life. One shop owner, in the midst of lighting candles to illuminate his shop, smiled a big toothy grin and burst into the opening chorus of "Happy birthday to you" as we passed by – as if to suggest that this blackout was entirely in our honour! Another sales technique?

As we wound up our purchases, we signalled to Ali that it was time to head back to the car waiting for us outside Bab al-Yemen. He cheerfully led us back, casually and politely brushing away several other would-be young guides along the way who tried to offer us their services. Ali was having none of this; "only guide," he would point at himself and say to us, to reconfirm his status. I admired the confidence of this young kid who had more street smarts than my well-travelled, worldly kids who've never had to help their family earn a living. As he helped load our bags into the car, we each handed him a 100 riyal tip (the equivalent of $1 US). With a polite smile, he thanked us and disappeared into the crowd back inside the city gate – returning to the shops to collect the rest of his commissions. Our car pulled away, leaving us to mull over the images of the souq that had been etched in our memories. Shopping in a suburban western mall will never be the same!!

Wandering re-establishes
the original harmony
which once existed between
man and the universe.

Anatole France

Photo courtesy of Ken Bowie – Masai

Falling in Love with Africa

Melissa Kooijman

My love affair with Africa began while studying Social Anthropology at university. One of my professors announced that if anyone was interested, he had learned of a program where students would camp, study and live in the African bush for three months (for credit!). Ready for an adventure I signed up! While travelling Kenya and Uganda, I gained first hand knowledge of the local cultures by interviewing Samburu and Masai warriors. I learned an enormous amount about the wildlife and while in Mombasa I participated in an archeological excavation and also successfully completed my diving certificate. My second trip was a one-month overland camping safari starting in Johannesburg and finishing in Nairobi. (The highlight from that trip was whitewater rafting and going on a three day canoe safari down the Zambezi River while the hippos and crocodiles looked on.) While returning from that trip I decided to change career and re-trained in the travel industry. In 1998 I started working with Wild Africa Safaris, owned by Chris Bradshaw. (Originally trained as a lawyer in his native Zimbabwe and re-trained in tourism when he relocated with his family to Canada.) My third trip was work related where I did site inspections at various hotels and lodges in Nairobi and throughout Tanzania and Zanzibar. While visiting Olduvai Gorge, I ran into three Masai boys who, instead of wearing the traditional red cloth, wore black instead with white painted faces. They had just been circumcised and were hunting lion to kill as part of their right of passage. I was told that if they killed a lion they would then become warriors and would get to wear the traditional red cloth. The boys did not speak any English so my driver guide translated for me. My fourth trip was a trip of a lifetime with my husband (his first time to Africa) where I did site inspections in Zambia, Botswana and the Seychelles. The Seychelles had the most incredible and stunning beaches I had ever seen. While snorkeling with my husband on La Digue with not another soul around, a sea turtle swam right by us and they got to swim with him. I also did a gorge swing right over the Zambezi gorge and I say.... it is a much bigger thrill then a bungee jump!

It gets in your blood and you just want more and more! I feel extremely fortunate to be doing work that I love and am very passionate about. "I fell in love with Africa on a familiarization trip to East Africa". I did about 32 site inspections in ten days, but in those ten days, I experienced the people and the culture. I visited Masai Villages where I was greeted by villagers singing and dancing and was invited to dance with them. I learned how to make a dung hut and shoot an arrow into the vein of a cow to collect the blood in a gourd so the Masai could drink it. I visited schools and brought over balls, pens, and books for the children who were so grateful. The entire school met in the field and sang a thank you song. I visited remote villages where the children had never seen a white person. They would come up to me and rub my skin, thinking that the white would rub off. I even went through a Masai marriage ceremony where I married a Masai warrior. We wore lion headdresses and danced around the fire. It was so amazing to experience such an incredible culture. Anyone travelling to Africa knows it gets into your blood! I could not imagine any other destination to represent and I will never stop going back.

Melissa Kooijman is with Wild Africa Safaris.

To the Roof Top of Africa: Mt. Kilimanjaro

Larry Henke & Dr. Jim Powell

Definition of courage: 'Grace under pressure' – Ernest Hemingway

It took only a couple of minutes. All of my senses were being bombarded with similar sentiments that had been lost to me from my first trip into the heart of this deep mysterious continent. All of the memories came roaring back to me. The smell of human sweat mixed with animals' odours and the dust from the surrounding land seemed to wrap its arms around me and welcome me back. It has been more than 20 years ago that I was here as part of a Trans African Safari and it was still as all capturing the second time.

It was early evening when we landed at the Kilimanjaro International Airport after a long 33 hour flight from my hometown Calgary. Four friends (Dr. Jim Powell, Ken and son Braden Bowie, Chad Labbe, my nephew) and I had planned this trip almost 10 months ago. Jim and I, while entertaining our wives at a Dinner Theater, chatted about our dream to travel to Tanzania and climb the highest mountain on the largest continent in the southern hemisphere, Mount Kilimanjaro or as the Swahilis called it "Shining Mountain." We were open to any of our friends joining us. Ken was the first, in his own words, "That's on my to do list." Braden was not about to let his dad go without him. Even though Braden was 15, he was in top condition and looked forward to the opportunity. Chad, my nephew from Winnipeg, (who is in his 20s and in his element with the outdoors), also jumped at the chance to travel with his 'Uncle Larry.'

As we were in line waiting to clear through the normally intriguing African customs and immigration, I glanced over at my friends trying to read their reactions to this foreign stimuli. I felt responsible for them, since I have been here before and had a lot to do with convincing them to come. All of their faces were of those of little boys entering into a candy store with strange but interesting treats for them to enjoy. After leaving the airport it was another short hour-long bus ride to our hotel and staging point for our short two weeks in Africa. Along the way we made a few stops at small roadside markets and were able to get a first glimpse at rural life. There were cars parked haphazardly around the market, people and domestic animals milling around and music blaring from behind the stalls. We were all too worn-out to make any comments on what we were seeing.

We arrived at the hotel, checked into our rooms, and then met in the outside bar for what we considered a well earned beer. The hotel is a small family owned business that caters to adventure travellers and operates as an outfitter for climbing Kilimanjaro. The hotel is similar to our motels at home, that is, the rooms around the outside of a square with a large open-air center courtyard for the dining area and patio setup with chairs and tables for socializing. The big difference is there were no vehicles in this inner courtyard and a few animals, cats, dogs, chickens, goats etc. seemed to have the run of the place. The beer hit the spot. It was cold and tasted great. The bottles were large and the cost was cheap. The names of the brands available were strange but fit into the romantic view of Africa. Tuscar, Kilimanjaro, Safari, to name a few. We powered out after a few drinks and retired to our rooms and listened to the sounds of the African night.

Next morning I looked to the north and there on the horizon poking out of the small blanket of clouds stood the subject of this trip, Mount Kilimanjaro towering 16,000 feet over the surrounding plains. The elevations at the foot of the mountain is less then 3,000 feet while the summit is 19,340 feet and in seven days we were going to climb to this summit and back down again. Now that I was looking at the real thing this was a very intimidating goal, but I was intent and determined to reach the "Roof of Africa". I spent the rest of the day with my group and toured the town of Moshi. We took in the sounds, sites, smells, tastes, and atmosphere of this small but hustling town. I was impressed with the changes that had taken place since I was last through here more than 20 years ago. There was an atmosphere of success and prosperity that was predominating, instead of the despair that I remembered from before. It seemed the government had recognized the power of the masses was better for the country, than the power and wealth of the few.

Day two in Africa started with unpacking and re-packing for the climb and meeting Filex, our guide for the next six days. We had our first lesson in scheduling according to African time. We were asked to be ready to meet our transport by 9:00 a.m. but the truck did not arrive until 11:00 a.m. This was a cue to unwind to 'African time' while simultaneously be fevered up for the climb. Not the easiest task since we all poured into our work the day before. Then our truck and the porter's truck broke down, not just once, but a couple of times. There was nothing to break the monotony of the wait until finally with the next passing truck we hitched a ride. So here we were a bunch of "explorers, mountain climbers, and adventurers", venturing on the journey of a lifetime in the back of a Toyota pickup. The truck made many stops in Moshi before we finally arrived at the National Park Gate five hours late and too late to proceed with the first day's hike. To paraphrase Pumba in The Lion King "Akuna ma tata". We resigned to set up camp at the park gate. With nothing to do but pass time, Chad and I took a short walk to an empty guesthouse. Chad found some suspicious looking samosa and began eating them. We both grabbed warm beer that was for sale, sat down and watched the sunset of the African plains that seemed everlasting. (Elevation at gate, 6,000 feet).

Morning broke bright and sunny and you could feel the warmth in the atmosphere. We had the morning meal of eggs, toast and jam, oatmeal, and instant coffee and then left camp by 8:00 a.m. to start our journey. Because we had not walked on Day 1, we needed to make up for lost time by reaching the planned second camp today. This would add about three or four hours to the planned six hour hike. Kilimanjaro is a mountain of contrast and extremes. You start from the tropical rainforest and finished in arctic conditions. The Rongai route we had chosen, approaches from the north side of the mountain and retains a sense of untouched wilderness compared to the well-travelled Marangu and Machame trails. We chose this route that was less travelled so we could tent every night instead of sleeping in A-frame shelters as offered on the other routes. The first couple of hours we passed through maize fields. Corn was growing in all different stages of growth. Intermixed with the maize were potatoes, carrots, onions, and other vegetables. There were local people living in these plantations so we stopped along the way and talked with them, giving the kids candies we'd brought with us for this purpose. As we walked up through this range, the temperature started to get hot and by the time we stopped for a rest at 10:00 a.m. it was already 34 degrees Celsius. I looked at the others in the group and we were all showing signs of excessive sweating. We all swallowed up water from our water bottles as if it was nectar from the gods. After a short break we set on upward until lunch. From here we had a superb view of Kibo and the Eastern ice fields on the crater rim. We stopped for only an hour and then walked steadily until 5:00 p.m. when we reached the second camp Kikelewa Moorland at 11,800 feet. We were all worn out from a long day and after a quiet supper called it a night by 8:00 p.m. Tenting in a four-man tent was a real challenge; we had our luggage, smelly boots, socks, and clothing hanging from tie straps. I envied Jim, for he had the one-man pup tent all to himself!

The third morning was cool to start with and after a couple of hours of walking through the giant Senecios forest and in and out of the fog, the sun broke out and the temperature started to rise again. This day was not as hot as the previous because we were getting higher up the mountainside. The climb along grassy slopes was a lot steeper today but shorter. By mid-afternoon we arrived at the base of the jagged peaks of Mawenzi (one of the volcanic peaks) and camp at Mawenzi Tarns at approximately 14,400 feet. Up to this point our water source was from clear mountain streams we walked past. After Mawenzi there will be no more fresh water and what was needed for the next two days we had to bring with us. The water here was dark green in colour and suspect at best. Before leaving home I had received a water purifier for Christmas from my wife and now it was going to be the most important tool we had. After straining the water through this purifier and treating with chemicals the water was safe to drink but not good tasting. We spent the afternoon resting and exploring the area as an aid to acclimatize. So far, the greatest challenge was the huge altitude gain.

Day four started off much cooler again and with a bright blue clear sky, we were eager to walk to the base of the mountain. The start of the walk was steep to get out of the campsite and then a long gentle slope cross the "Saddle" between Mawenzi Peak and the main peak of Kilimanjaro. The "Saddle" looked like the lunar landscape with sand, rocks, and large boulders strewn about. We arrived at Kibo Hut, 15,000 feet, mid afternoon after four to five hours of walking. We were feeling the affects of the high elevation and found it difficult to catch our breaths. After lunch the remainder of the day was spent resting and gathering our strength and energy for the final push to the summit of the mountain. That evening our supper consisted of leftover tuna casserole. The combination of hard-boiled eggs, slices of bread and apple jerry and hot Milo (chocolate) did not agree with our palates. I thought this meal was to be the most essential meal of the walk and even though I was disappointed, devoured what I could. During supper we had a short meeting with our guide whereupon he informed us of what to expect when we started out at midnight. Our preparations included four layers of clothes, toque, mitts, and a walking stick. He said it would be cold when we started and gets extremely cold as near the summit. Coming from Calgary and Winnipeg where it is normal for temperatures to dip down to minus 30 degrees Celsius, the Siberian warning did not alarm us. We turned in early after supper and tried to sleep. My mind was agitated thinking about the attempt to reach the summit; hoping no one would

get altitude sickness and that we would descend safely. I thought of an ancient tale that tells of the local Chagga who believed in Gods living on top of Mt. Kilimanjaro. They tried to steal the white gold, only to be struck down with headache, nausea and cold. Now we know of altitude sickness and have discovered that the white gold was actually glacial ice. To avoid the same problems as the early Chagga climbers, we all had to ascend very slowly.

I wasn't sure if I fell asleep or not, until our guides woke me at 11:00 p.m. We had to get up and get moving. Half an hour later we began the final and by far the steepest and most demanding part of the climb by torchlight and what little light the stars and moon illuminated. The pace the guide set was deliberately slow and we started to zigzag. Slowly, in the dark up on a switchback trail we shuffled through loose volcanic scree. We all felt more like a chain gang and hence it was either Ken or Chad that coined the simultaneous strut 'The Kili Shuffle.' Within ten minutes of starting, I was puffing like a steam engine trying to catch my breath. I thought to myself "how was I going to complete this climb if I feel this out of breath already?!" I started to count my steps to keep in time to my breathing. This took my mind off the agonizing pain and I settled into the pace of the climb. Soon we were half way up so we rested and lapped in the luxury of the view. It was a spectacular sight. As we climbed under a clear sky with a full moon and a blanket of scintillating stars, we could see the short strings of the lights from other groups of climbers coming up the path behind us. As we pushed further up the mountain the temperature started to drop and the wind increased. After five and a half hours of intense trekking and excruciating pain we

Ken & son Braden Bowie from the top of Mt. Kilimanjaro.

broke over the top of the rim of the crater, Gillman's Point 18,500 feet, the first summit of Kilimanjaro. It was still very dark, the moon had set, and the wind was sheer cold and brisk. We stopped for five minutes to take in the situation and then pressed on to the final summit. The slope of the path was not steep but the effort it took to continue was extreme. Now it was a mind game to continue on as the lack of oxygen has fogged our judgement. I was always convincing myself it was worth the effort to reach the final summit. As we proceeded westward, we needed to concentrate on the feet of the person in front of us. Concentrate that was the key… soon someone said that their headlamp was getting dim and we need to stop to change the batteries. We then noticed all the headlamps were dim and we felt a shot of anxiety because no one thought the spare batteries were going to be required. We stopped, turned around and noticed the sun was starting to rise and it was the light of the sun making it appear like our lights were dimming. Just one of the effects high altitudes has on your power of reasoning.

It took us another 30 to 40 minutes for us to reach the rooftop of Africa, the highest point on earth to which you can walk. Uhuru Peak 19,340 feet (5,895 metres). We were elated, exhausted and in awe of the magnificent glacier and ice cliffs that still occupy the summit area. When I was here approximately twenty years ago, glacier and snow covered the entire summit. Now the ice shield covered a scant one third area of what I had previously observed.

We spent ten to fifteen minutes at the summit hugging, congratulating each other, and taking our pictures under the summit sign. We looked around at the splendours of Africa that were unfolding all around us. Ice and fire, a reminder of what formed Kilimanjaro as we looked down into the cauldron of an extinct volcano. We were so high up you could see the curvature of the earth and feel the lower jet stream winds touching, which was blustery and unsympathetic. After a brief period we turned around and started down. However, before we got far I remembered the satellite phone I packed. We stopped and took shelter behind a large bolder and each of us called home. I called my wife 'twice'; with all the heavy breathing she thought I was an obscene phone call! We were all choked up as we spoke to our families at home and heard in their voices their joy, pride, and concern in what we had just accomplished.

It took us close to eight hours to climb 4,300 feet and less then three hours to descend back to Kibo Hut for a much-needed rest and some energy food. After three or four hours rest, we were off again descending to the final campsite at Horombo Hut. We indulged in a cold beer or two, some refreshments and a long nights rest. Some enterprising individuals had packed up beer for sale two-thirds of the way up the mountain. Eight hours of steady descent took us through moorland to Mandara Hut, on the Marangu route. (The route I took 20 years ago, has not changed.) We continued down to the National Park Gate and back to the original hotel we began at. The adventure of a lifetime was over as we all received our certificates for reaching the summit. But our trip to Tanzania was not over, the best way to get over a trip was to book another trip, so we began our weeklong safari for some much needed R & R.

Ras al-Hanut

Morocco, Algeria, and Tunisia form what is known as 'The Maghreb.' This coastal plain is very fertile and produces cereals, citrus fruits and vegetables. The marrying of fruit with meat requires the expert use of spices to enhance the altogether exotic flavours of curry, saffron, paprika and ginger!

A Recipe for Ras al –Hanut
2 teaspoons ground cinnamon
1 teaspoon turmeric
½ teaspoon freshly ground black pepper
¼ teaspoon freshly ground nutmeg
¼ teaspoon freshly ground clove

The trick is getting the proportions right. Store in a jar.

Over-the-Counter Encounter: Pokhara, Nepal

Steve Clifford

It was early October and I was enjoying some R & R in Pokhara, a small town in the shadow of Nepal's Annapurna Mountains. I was due back in Kathmandu on the weekend, where I was to join a small group on a trek to Everest Base Camp. I needed a new daypack and decided to check out the huge variety of trekking shops that lines both sides of Pokhar's main street.

An hour and half of searching and haggling had produced no results and I was mystified as to why this was taking so long. Having spent five months in the Middle East and South Asia, I knew the routine pretty well. Nevertheless, I found myself standing in the dusty street, hands on hips, another perplexed westerner trying to solve a "problem". I eventually snapped out of my stupor, determined to buy a bag from the next shop I saw.

That tiny shop was stuffed with trekking supplies hanging from the walls and ceiling, and piled high in the corners. Behind a small wooden counter stood a young man, lighting up the poorly lit shop with his beaming Nepali smile. We shook hands and within 60 seconds I had purchased the perfect bag at the perfect price.

As I handed over my money, I noticed he was wearing a small Canadian flag pin, not an unusual sight as they are often given as tokens of appreciation from Canadian travellers. I pointed out the pin and he quickly replied that it was a gift from friends in Canada. I assumed from people he met in his shop but he corrected me, saying he had led them on an Annapurna trek several years ago. And that's where it should have ended. But I felt compelled to continue my questioning, both of us equally puzzled as to why.

"Do you remember their names?" I blurted out. He thought carefully and replied, "Martin, he was tall and worked with computers..." He noticed the stunned look on my face as I said "and Brian gave you the pin". He nodded and the shared sense of disbelief intensified when I explained to him that Brian was my roommate back in Canada. His smile now filled the entire shop. He introduced himself as Khrishna and quickly led me across the street and into another store, eagerly pointing out the photograph of himself with Brian, Martin and several others enjoying the hot springs at Tatopani. We made plans to meet for tea the next day, where he would introduce me to his newlywed wife.

I returned at the appointed time, and as we sipped our tea he produced a letter he wanted delivered to Brian, as well as a small photo album given to him by the trekkers. As I flipped through the album of stunning mountain scenery, I found my jaw dropping for the second time in two days. Staring up at me from this tiny album in a tiny shop at the edge of the Himalayas was a photograph of my home in Victoria, British Columbia.

After enjoying our tea, we snapped a photo of him and me in front of his shop, which I later included in the letter to Brian. Ten years have passed and I never did meet up with Khrishna again, despite attempts on a return visit several years later. Brian and I still talk about that fateful encounter in dusty Pokhara – and my "lucky" daypack continues to accompany me around the world.

Steve Clifford is with G.A.P. Adventures.

Climbers and Sherpas in Nepal.

Pedalling to Nirvana: Mount Kailash by Bicycle

Graydon Hazenberg

"The highest road in the world." The words grabbed my attention when I first read them. The road being described was the Xinjiang-Tibet highway in western China, and like a fool, I thought this was an invitation to ride a bicycle along it.

Graydon and Kailash Children

Along with my sisters Audie and Saakje, I had done my fair share of travelling, including bike travelling, but this sounded like a real adventure, the sort of out-of-ordinary experience that we were craving for our first trip together since family vacations by car back in high school.

This trip seemed to have all the necessary elements for an adventure. For one thing, it was in a remote corner of the globe that almost no Westerner ever travels. For another, it was going to be physically difficult and demanding. The scenery promised to be spectacular. After all, we were going to be crossing the roof of the world, some of the highest mountain ranges in the world. The road was officially off-limits to foreigners, so there was the additional thrill of being naughty and doing something forbidden. And finally, the goal, far-off Lhasa, the Forbidden City, was alluring enough that the tremendous effort of getting there would be worth it.

The winter before the trip, stuck in a gray and almost snowless Toronto, I read as much as I could about the area and about the history of Western exploration of it. I read about adventurers like Nikolai Przewalski, Alexandra Stevenson, Francis Younghusband and Ewai Kawaguchi, all trying to get to Lhasa in the face of Tibetan hostility and inhospitable nature. Audie and Saakje, along with their respective partners Serge and Luke, agreed that we would be worthy successors to these hardy explorers.

As a warm-up, we spent a month and a half cycling and hiking through the mountains of the Himalayas, Hindu Kush and Pamir ranges on our way from Islamabad, Pakistan to Kashgar, Xinjiang, in the far western corner of China. After 2,000 kilometres of riding, 400 kilometres of it over dirt roads and jeep tracks in unimaginably poor condition, we felt we were ready, physically and psychologically, for the second and more challenging half of the voyage, that is, biking from Kashgar to Lhasa, 2,800 kilometres away.

After 22 broken spokes, ten snapped bolts, two broken luggage racks, two shredded outer tires, innumerable punctures, a shattered bottom bracket and an unfortunate episode with a bus that left one of Saakje's handlebars shorter than the other, we felt that nothing could come our way that we hadn't already overcome. Even the various stomach ailments that had affected us all seemed to have cleared up by the time we left Kashgar.

Leaving Kashgar was difficult; we had grown used to a steady diet of Uighur bagels and cream cheese, Szechuan noodles and innumerable ice creams as we tried to fatten ourselves up for the lean kilometres ahead. The Sunday Market, a vast sea of 100,000 Central Asian faces in distinctive hats and headdresses, was such a photographer's paradise that we were tempted to stay another week to see it again. In the end, however, we packed our bike panniers full of instant noodles, tea, raisins, nuts and tins of fruit and set off.

The riding on the first few days was easy enough, flat and hot, skirting the Taklamakan Desert. Our only problems came with the vast and unruly crowds that we attracted whenever we stopped. Hundreds of curious locals would gather around us, hemming us in on all sides, and soon adventurous hands would dart out, trying to help themselves to whatever was loose. Accustomed as we were to attracting crowds in Pakistan, this was much worse and our tempers were quickly frayed. We ended up fleeing the oasis cities of Yenihisar and Yarkand as quickly as humanly possible; just to escape the huge, unruly mass of staring eyes.

On our third day out of Kashgar, we reached Kargalik, where the road to Tibet branched off to the south. Our propensity for drawing crowds landed us in hot water (or, more accurately, cool water), as the local police easily got wind of a raggle-taggle band of bicycle gypsies in town. Mr. Ahmed Chan, of the Public Security Bureau, gave us a stern lecture on the consequences of trying to cycle to Tibet, over a road that was closed to foreigners.

"If you try, I hear about. I come in truck pick up you, back to Kashgar, take passport, pay fine, back to you country."

We protested that we were going in the other direction, to the east, and that we would never dream of breaking any of the laws of the People's Republic. Then we went and hid in the sand dunes outside town to plot our next step.

Luke was pessimistic, worried about Ahmed Chan making good on his threats, and was half-convinced that our trip was over before it had begun. Saakje and I were the most positive about pressing on, albeit cautiously. We eventually decided to wait a day to let our trail grow cold, and then to sneak onto the road to Tibet under cover of darkness. We succeeded, but only after being followed in the dark by a car without headlights, which made us abandon our first attempt. On our second try, we spent a tense day hiding near, and occasionally in, an irrigation canal, watching Ahmed Chan's Land Cruiser looking fruitlessly for us in the wrong place before he gave up the search.

We breathed a large sigh of relief, and then spent the entire night cycling in the dark, with no lights other than the distant, eerily flickering orange glow of distant oil wells flaring off natural gas. When we awoke after a morning's exhausted sleep, we found that the landscape had changed in the dark. Gone were the flat desert plains of Kashgar and Kargalik, and in their place rose dark, barren mountains, the foothills of the Kun Lun range, whose name means "mountains of darkness." We speculated that the name comes from almost total lack of snow on the peaks to lighten the dark, gloomy rock faces.

We started cycling by day now, convinced that it would not be worth Ahmed Chan's while to come this far in pursuit of us. From that point on, our days fell into a routine. Saakje would set her alarm for 6:00 a.m. and try to rouse the rest of us. We would lie in our tents, trying to convince ourselves that it was worth getting up. Breakfast would be raisins, peanuts, and tea – adding bread and hard-boiled eggs if we had passed a truck stop recently. Finally, around 9:00 a.m., having taken down camp, we would set off with the intention of covering 70 kilometres.

Before this trip, Saakje, Audie and I would have scoffed at such a short daily distance, since on other bike trips we were used to cycling at least 100 kilometres a day. On dirt roads, however, with our panniers bulging with tents, sleeping bags, stoves, cold-weather clothing, heavy spare parts, tools and kilograms of food, we simply could not force ourselves to go further.

Any hint of wash boarding sent our panniers relentlessly bouncing on our luggage racks, eventually snapping screws or even the rack itself. We had one spare rear rack, and once that was used up, a broken rack would mean the end of someone's cycling trip. Thus we kept our speed low, grimly bumping along, heads down to watch for potholes or large rocks on the road, watching the kilometre posts and the telephone poles creep by at an agonizingly slow pace.

We would stop to refill water bottles whenever we passed decent looking water, which was infrequently. We treated our water with iodine, and sometimes Luke or Serge would pump water through their ceramic filters, although this was a glacially slow process and the pumps had a tendency to clog up completely with the silt in the water. No matter how much we tried, it was impossible to avoid occasional intestinal problems, given how filthy some of the water was.

Eventually one or all of us would have had enough of riding, and we would stop for an unsatisfying lunch of more raisins and peanuts. We were reluctant to break out our stoves and cook noodles, since it took so long, and we had little else in the way of lunch food. We tried making chapattis the night before, but it was a long process that usually resulted in tiny fragments of bread, burnt on the outside and still raw on the inside. Then we would lie beside our bikes, summoning the will to continue, before setting off again.

For several hours, we would cycle along the deserted road. Our conversation revolved around food. Audie and Saakje, in particular, were obsessed by visions of cheese, and a frequent afternoon conversation – starter was "What would you like to eat right now?"

Finally we would give up the struggle, towards sundown, usually short of our 70 kilometre goal: only twice in the coming 25 days of cycling did we top 70 kilometres, usually settling for something in the mid-60s. The last few kilometres would be spent searching for a good camping spot, where we would set up our tents, cook instant noodles, write in our journals and collapse, exhausted!

Most evenings would see one or more of us fighting with a malfunctioning stove, cursing MSR (the manufacturer) as we cleaned and blew through fuel lines and prayed to the gods of cooking! We carried along a small guitar, a chessboard and playing cards, and yet we never seemed to have the time or energy to play any of them. Our spirits seemed to be broken by the slow, bumpy cycling, and our bodies were being weakened by the lack of adequate food.

My body, more than everyone else's, needed food because of my abdominal fellow travellers. I had become ill with stomach problems in Pakistan, as had all of us except Serge. Unlike the others, however, my dysentery refused to leave once its welcome was over. I had been in agony with cramps, bloating, diarrhea, and crippling heartburn on the way to Kashgar, and although the symptoms would disappear every time I dosed myself with Flagyl, they would reappear as soon as I stopped. I started to lose weight at an alarming rate; everyone else was getting skinnier as we failed to feed ourselves sufficiently, but I was the one becoming the most skeletal.

As we were making our final push up the pass, through beautifully green high-altitude meadows, Luke's long-suffering back luggage rack finally gave up the struggle. Luckily, we were carrying one spare back rack, since otherwise Luke's trip would have been over. It took a long time to install the new rack and adjust the panniers to fit on it. Meanwhile, precious daylight hours slipped by.

By the time we reached the summit, the sun had set and it was freezing cold. Luke was exhausted, Saakje was suffering terribly from cold in her fingers, and she and Audie were both feeling altitude sickness. What should have been a triumph, our first 5,000 metre pass, instead became a grim descent in the dark. Fully clothed, we searched for a place to camp. We ended up camping beside the road, on a flat patch of ground that turned out to be a truck driver's garbage dump, littered with truck parts, smashed bottles and dung. It snowed on us overnight, and we awoke feeling like death. The tough decent warmed the girls and us up and we felt much better the further we descended.

At the foot of the pass we gorged ourselves in a miserable noodle shop for truckers that lay at the junction with the road to K2 Base Camp. We couldn't see the second-highest mountain in the world, but further up the valley we met a convoy of camels that were headed for Base Camp to work as baggage animals for a mountain-climbing expedition.

We decided to be sensible about the next 5,000 metre pass, the Keycha Daban, and to take it slowly. We camped at the bottom of the steep climb and set off early, stopping every five kilometre and cycling very slowly to conserve energy. A convoy of Chinese army trucks was grinding down the pass as we were coming up, and several soldiers leapt out to take pictures of them with us.

We reached the top sooner than expected, and set off excitedly to try to reach Shahidulla, reputedly 50 kilometres away and all downhill. We encouraged ourselves with shouts of "Noodles and beer in Shahidulla!" Of course, this trip being as it was, we couldn't make it 50 kilometres downhill despite trying all afternoon. The road surface went to hell, and unfriendly headwinds stopped us nearly dead in our tracks.

The cycling, though, was grimmer than usual, and we ended up abandoning the road for long stretches in search of a smoother ride across the desert surface. We ran out of water several times and were reduced to begging water from passing trucks. Word seemed to spread along the trucker's grapevine that we were in trouble, as several trucks stopped (without asking them to) to offer us water, soft drinks, and the best-tasting watermelons we had ever tasted.

Audie was perpetually cold, unable to sleep at night because of shivering, and all of us were cycling in long sleeves, long tights and jackets. My dysentery continued to get worse. Our tents were buffeted by winds, and one particularly windy morning only a flying tackle by Luke saved his tent from blowing all the way to Lhasa.

After four days and 180 kilometres, we were glad to climb a low, rounded knoll that was the highest point of the trip, a nameless pass at nearly 5,500 metres, and find the telephone poles on top decorated with Tibetan prayer flags. We had finally arrived in the Promised Land, and the excitement of getting to Tibet momentarily dispelled the gloom about cycling conditions.

The border between Xinjiang and Tibet was more than just an arbitrary line on the map; there was a sharp distinction between the dusty, dead landscape on the Xinjiang side and the luminescent green grasslands in Tibet. There was more life in the landscape, with occasional sightings of fox, and gazelle, and eagles.

Here and there, in the distant patches of green, we could see the black felt tents of the drongpa, the hardy nomads of Western Tibet. They spend their summers grazing their herds of sheep and yaks in the high altitude meadows at 5,000 metres. We stopped in at one tent, surrounded by scraps of sheared wool, smashed beer bottles and yak stomachs full of butter, and met the women of the family. They were dressed in their traditional hand-woven clothes, adorned with silver jewellry, and their weathered faces showed how unimaginably tough their existence was.

We spent the next six days plodding towards Ali, the less said the better. As we descended, we lost our beautiful high-altitude greenery and the road, if possible, deteriorated yet further into a series of parallel, hopelessly wash boarded tire tracks. At times, outside Ali, there was so much sand on the road that we had to push our bikes for kilometres at a time. Our tempers frayed and Saakje and I nearly punched each other arguing about a broken pump. More equipment broke, that is, pumps, pedals, and pieces of luggage racks. We crashed a few more times. We also, unanimously, decided to abandon hope of reaching Lhasa by bicycle, given how slowly we were moving, the horrible road conditions, and the fearsome river fordings said to lay beyond Mt. Kailash. We decided to try to get a lift to Lhasa in a truck or a Land Cruiser once we had reached Kailash.

Three things in particular stand out in my memory. When we arrived in the town of Domar we spent blissful hours stuffing our faces, and drinking beer, and playing cards in a restaurant. We were relieved to have real food again. Past Domar, we rode along the barren shores of long, narrow deeply blue Lake Panggong, stopping now and again to swim. And 100 kilometres before Ali, we found Buddhist and pre-Buddhist petroglyphs beside the road and spent a happy hour photographing and sketching them.

Ali was, perhaps, the greatest disappointment of the entire trip. We had read about hot showers and creature comforts, and even contemplated spending a rest and recovery day there. It was a remarkable town, a little piece of modernizing China, complete with paved roads, taxis, and multi-story concrete monstrosities, plunked down in the sandy, wide valley of the upper Indus River.

When we arrived, however, we couldn't find the hotel mentioned in our guidebook, and while searching for another one, we were arrested and had our passports confiscated. We each had to pay forty dollars to get them back, which was letting us off lightly for having travelled across a closed road. Our hotel, when we found one, not only had no hot showers, it had no running water of any kind, and was upstairs from the only disco in Western Tibet. We were kept awake much of the night by earsplitting music, the roar of an electric generator and a remarkably prolonged drunken brawl!

Having eaten three enormous meals in 18 hours, we fled the next afternoon, determined to get to Kailash and end the misery of cycling. However, it would take us eight days to cover the last 300 kilometres as all Four Horsemen of the Cycling Apocalypse came to haunt us, that is, wash boarding, loose sand, enormous boulders, and river crossings.

By the end of the second day of cycling, Luke and Serge had decided to quit and hop onto the next truck to Kailash. As fate would have it, though, not a single vehicle passed us all day, and they had to keep riding.

On one memorable day we covered 33 kilometres and took ten hours to do it, as we spent much of the day floundering in river crossings. The Zanskar Mountains on the border with India lay to our north, and the monsoon rains that were falling on them spilled across our road in an endless series of side streams. One river had more than 70 separate channels that we had to ride through, push through or, more frequently, carry our bikes across. We were soaked to the skin and frozen by icy storm winds. Saakje and Serge were both nearly swept away by raging torrents. It was, in Luke's words, "a terrible suffering business." My dysentery attacks were becoming more frequent and more severe. And to top it off, the local dogs all took a disliking to me and I had to keep them at bay by hurling rocks at them.

It was sheer force of will that brought us to Moincer, where we planned to visit Tirthapuri Monastery, one of the main pilgrimage sites of Western Tibet. There we found Bruce and Jackie, two New Zealanders whom we had met in Kashgar. Hitching lifts in trucks, they had just arrived. It had taken them almost four weeks, just as long as it had taken us. They had spent two weeks waiting for one lift. It cheered us up slightly to think that we hadn't lost any time by going by bicycle.

Tirthapuri was wonderful, a monastery thronged with pilgrims, full of butter lamps and prayer wheels and a sacred geyser. It was a relief to have arrived somewhere and to experience some culture at long last. I took exploratory spins on my bicycle, having taken off the entire luggage, while the others took to their feet to escape the hated bikes. That evening, while I listened to two monks chanting on a hilltop, I followed their gaze and saw on the horizon Mt. Kailash, our destination. It seemed incredible that we could finally see the end of the road.

It still took us two days and our scariest river crossing yet to reach the enormous tent city of pilgrims at Darchen, at the foot of Mt. Kailash. The scenery became more and more grand, with huge, sweeping vistas over a wide plain over to the Himalayas, on the Indian and Nepalese borders. In front of us shimmered a turquoise lake, Raksas Tal, and unseen behind it was Lake Manasarovar, the most sacred lake in Tibet.

A few groups of Western tourists, who had paid thousands of dollars to come to Kailash, were scattered amongst the thousands of Tibetan pilgrims at Darchen. There were even a few middle-class Indian tourists, as Kailash is sacred to Hindus, who regard it as the home of the god Shiva. A party of Italian tourists burst into applause as we arrived, bedraggled and grimy and emaciated.

As it is with all sacred sites in Tibet, the greatest spiritual benefit is to be gained by walking around Kailash, a two-day process (although some determined Tibetans race around in one long day). The scenery was fantastic, with golden-topped monasteries dotting the landscape and sheer valley walls that sprouted instant waterfalls whenever we were drenched by a cloudburst. Kailash kept its zebra-striped upper reaches shrouded in cloud most of the time, but we awoke the second morning to clear unobstructed views of the sheer north face.

There was a steady stream of Tibetan pilgrims wearing black pilgrimage robes, swinging prayer wheels, the women carrying babies on their backs, some of them accompanied by their yaks. We got lost somehow and crossed the wrong pass, but we prostrated ourselves at all of the prostration points, so I hope that we earned enough karmic brownie points to save our souls. It was such a relief to have made it to Kailash and to get off our bikes that we were sorry when we finally trudged back into Darchen.

Now only two things remained to be done. We had to get to Lake Manasarovar, and then we had to get back to Lhasa or Kathmandu. Manasarovar is the lake that was once rumoured to be the source of four great rivers of Asia, the Ganges, the Brahmaputra, the Indus and the Sutlej. In fact, only the Sutlej starts at the nearby Raksas Tal, but the other three rivers have their sources within 50 kilometre, and in some sense, the Kailash-Manasarovar region may be said to be the navel of Asia. It lay only 30 kilometre from Darchen, but it still took nearly a day to cycle to it. When we got to it, we were so happy that we threw our bicycles into the lake to bathe them in the sacred waters, and to celebrate the end of cycling. They emerged covered in green slime. We rested up in the monastery guesthouse, bathed for the first time in a month in the local hot springs, and waited for our lift back to civilization.

When we got into our little convoy of four Land Cruisers and a five ton truck, we thought our troubles were over. That first morning, we sat in our Land Cruiser, watching the scenery fly by, playing guitar and chess in the back seat. Reality soon set in, however, and we spent five long, hard days driving the 600 kilometres to Lhatse.

Our truck got stuck in a river and it took five hours to pull it out. We got mired in mud numerable times. We ran out of fuel. We were denied permission to cross a newly built bridge, and had to take a 100 kilometre detour in search of a crossing point. Our bicycles were battered to pieces in the back of the truck. We ran short of food. I was crippled by my worst attack yet of dysentery. We drove 16 hours a day. And, on the last day, one of the other passengers, a young Chinese woman in another Land Cruiser, died of altitude sickness and arrived in Lhatse sewn up in a sheet in the back of one of the vehicles.

In Lhatse, we rejoined the Friendship Highway, the main road linking Lhasa to Kathmandu. I was almost late for my job, so I headed directly for Kathmandu, while Audie, Saakje, Luke, and Serge headed on to Lhasa with our convoy. I begged a lift with a Land Cruiser full of Polish tourists, and we raced down to near the Nepalese border, where landslides stopped us across the road. I got up early the next morning and cycled down to the border, carrying my bike across the landslides. The fact that my bike seat had been lost in the truck did nothing for the comfort of the ride. Within a 20 kilometre stretch, I left the arid, high-altitude world of Tibet and entered hot, humid, monsoon-soaked Nepal, where I cajoled and bought my way onto the next flight to Europe.

The others spent two weeks relaxing and rebuilding their strength in Lhasa, prowling around the temples of that mythical city, before flying to Kathmandu. I was disappointed not to have made it to Lhasa in the end, but glad that we had all survived. Looking back on the trip from the comfortable, well-fed world of hindsight, it was an amazing experience, and an unparalleled chance to see a remote corner of the globe, and it was satisfying to think that we had successfully cycled the highest road in the world. The huge vistas, crystal-clear visibility and amazing mountain landscapes that we had seen were irreplaceable experiences. It was wonderful to see Tibetan culture up close, and to have done the circuit of Mt. Kailash.

Nothing could have improved the slow pace the jarring riding, the head-down hard work on these appalling so-called roads involved… perhaps next time we'll take horses?

Graydon Hazenberg is from Thunder Bay, Ontario. He is currently travelling the world.

I dislike feeling at home
when I am abroad.

George Bernard Shaw

Dri Ving Miss Dei Xi

Dr. Hisham Zerriffi

After spending a few days in a car on some pretty bad roads in Hubei province and observing Chinese drivers in both cities and small towns, I have come up with some general rules of driving in China:

- Lanes are merely suggestions.
- Yellow lines are merely strong suggestions and can be disregarded in order to pass slower vehicles (true for both urban and rural settings).
- Use of the horn is mandatory for communicating with other drivers.
- Communicate with other drivers as often as possible.

Our driver, in addition to observing the general rules stated above, also seemed to have his own set of very interesting driving rules:

1. Unless absolutely necessary, avoid use of first gear.
2. Always drive one gear higher than you should in order to ensure sluggish acceleration until reaching highway speed.
3. Once highway speed is obtained, drive like Mario Andretti on drugs.
4. Always pass trucks on mountain roads when on a blind curve.
5. Religiously obey Rule 2 when passing on mountain roads to ensure difficulty in accelerating out of any difficult situation if a car comes the other way.
6. Ignore Rule 2 if the road is full of potholes. In that case, attack with reckless abandon.
7. Under no circumstances should one check for traffic before merging or use their mirrors when changing lanes.

Observing this unique driving style, particularly when he was observing Rule 7, made me wonder. Perhaps this is some manifestation of Buddhist philosophy. Something to do with both the interconnected nature of all creation and the impermanence of our existence? This has led me to the following: "A theory on the religious origination of driving behavior patterns" (I am hoping for some National Science Foundation funding to conduct a proper empirical study). Here then is the basis of my theory. With the exception perhaps of American, Canadian and British drivers, a similar phenomenon can be observed worldwide: Drivers appear absolutely fearless (or reckless) and by any measure, absolutely dangerous. This is not just Delhi or Beijing, think of the reputation of Italian drivers. However, while the observed phenomenon may seem the same on the surface, I believe it is a manifestation of very different world views. Here then are some of the religious origins of dangerous driving worldwide for some of the major groups:

Hindus: We will all be reincarnated so what is there to fear?

Muslims: It is all in the will of God whether we make it to our destination alive.

Catholics (e.g. Italians and Latin Americans): Aggressive driving is a creative outlet for pent-up guilt and frustration.

Buddhists: One must stop being attached to worldly concerns such as lines on the road and the truck coming down the other way as you try to pass on a narrow mountain road.

Protestants: (e.g. Brits and majority Americans): Their generally reserved nature (particularly the British variety), make them the exception to the crazy driving rule and thus proves the point.

Dr. Hisham Zerriffi is a visiting Fellow at the World Resources Institute at Stanford University.

In Zanadu di KUBLA KHAN
A stately pleasure-dome decree:
Where ALPH, the sacred river, ran
Through caverns measureless to man
Down to a sunless sea.
So twice six miles of fertile ground
With walls and towers were girdled round:
And there were gardens bright with
sinuous rills,
Where blossomed many an incense-
bearing tree;
And here were forests ancient as the hills,
Enfolding sunny spots of greenery.

Samuel Taylor Coleridge
...after reading Travels of Marco Polo

Celebrating Chinese New Year's in Hong Kong

Cecilia Lo

Just as we disembarked on our cruise in the spectacular island of New Zealand, we rushed over to the airport to catch our connector flight to Hong Kong. Even though I have been to Hong Kong several times since moving to Canada, I have not had the pleasure of enjoying Chinese New Year in Kowloon since I was a teen.

On December 29th, we landed at Chek Lap Kok Airport (also known as Hong Kong International). You could sense the anticipation of the festive occasion of Chinese New Years everywhere in the atmosphere, on the streets and in the smiles of strangers passing by.

Chinese New Year's begins with the new moon on the first day of the New Year and ends on the full moon 15 days later. We stayed at my brother-in-law's house, which only added joy to the occasion, as New Year's Eve is a family affair, a time of reunion with relatives arriving from all over the world to join with their extended family. Before we share in a festive dinner, we all burned incense in honour of Heaven and Earth, the gods of household, family and ancestors. We ask them to protect us from evil and illness, to guide us and bless us with good fortune. Traditionally, two red blessing verses hang on both sides of the door. The colour red is to chase away evil spirits from entering the house and as an offering to the gods that guard the door, a plate of fruit and burning incense is placed on the doorstep. As children we use to light up firecrackers to frighten the monsters away and also to welcome the New Year. This tradition has since been banned in Hong Kong due to too many house fires! However it is still allowed on Mainland China. The celebration continues on the streets with an enormous parade of decorative floats, bands and cheerleaders that would rival that of the Calgary Stampede Parade. The market was filled with scents of fresh florals trailing as far as your eyes could see; orchids, mums, and of course the famous little oranges or mandarins. The mandarins have been cultivated so the fruits are grown in clusters hanging on the branches. An abundance of oranges means a fruitful new year.

The first day of the lunar New Year is "the welcoming of the gods of the heavens and earth." All the shops are closed and it is customary to wear your finest new clothes. Since it is traditional to refrain from visiting family and relatives on this day we went to the mall! Some of the larger malls provided entertainment including the famous lion dance. The strongest martial arts students perform the lion dance. We watched along side with my little nephew who was skirting around me as the music grew louder and louder and the lions became more ferocious in his eyes! The lions went from shop to shop, sure not to miss any of the lettuce that was hanging for them to devour. Most people abstained from eating meat on this first day, as it is believed that this will ensure longevity and happiness. Vast amount of traditional food was prepared for us, such as a vegetarian dish called jai, made out of root vegetables. Coincidentally, the price of vegetables would also increase, almost doubling in cost!

The second day, we visited friends and relatives and got caught up in the usual tradition of a family banquet table. Together as the living, we prayed to the spirits of the ancestors, celebrating as one great community. Also known as "surrounding the stove" or weilu, it symbolizes family unity and honours the past and the present generation. You never arrive empty handed, so we brought fruit, pastries and Lai Sea in red envelopes for all the children in sight. On this day, besides the traditional root or fibrous vegetables, we were able to indulge in chicken, which is symbolic of prosperity. The chicken must be presented with its head, tail and feet to symbolize completeness. A whole fish was included in the banquet to represent togetherness and abundance. Noodles, uncut of course, stand for long life. The tremendous amount of food combined represents abundance and wealth for the households.

We drove to Wong Tai Sin Temple, in the heart of urban Kowloon, where at the beginning of the year, if you make an earnest plea to Wong Tai Sin, your wish will come true during the year under his blessing. The temple is an important religious center and the beautiful scenic ornaments surrounding the traditional buildings attract visitors from all over the world. It is located in a tranquil natural setting that is remote from the nearby housing estates and the bustle of the streets. Inside and outside of the temple, big urns are placed for the believers to burn incense and made their wishes. After the temple, we drove to another spot for a second New Year wish. It is the wishing tree. The wishing tree is a huge tree with several huge branches extending on all sides. We would tie an orange with red string on one side and our wish written on a piece of paper on the other side. We threw our wishes towards the branches hoping they would catch and hang on, as it is believed if your wish hangs on to the tree then it is supposed to come true. The tree is filled with wishes, which is a real sight to see.

On the third day, the festivity continues, although a little quieter, we went to a concert. On this day son-in-laws are also to pay respect to their parents-in-law. There's still energy excitement and celebration in the atmosphere as shops begin to open and routine sets in. However, I am sure more food is consumed during the 15 day celebration than any other time of the year! (The 15th day is called Lantern Festival, which is celebrated at night with lantern displays and children parading around with lanterns in hand.)

A Tokyo Canadian

By David Lindsay, President, Tokyo Canadians Hockey Club

Being the only gaijin (foreigner) on a Japanese men's league hockey team in Tokyo can be a bit frustrating. There is no lack of enthusiasm amongst the Japanese players. However, their general hockey-sense and play-making skills are not always the greatest. Above all, communication can be a real problem.

After playing in Japan for a while, one starts to pick up on the local hockey lingo. A goalie is called a "keeper," so after making a save it is common practice for his teammates to yell, "Nice keep, nice keep!" A shot, although not usually too powerful when taken by the Japanese, is typically congratulated by "Goodo shoot-toe!" My favourite is the Japanese name for the dreaded too-many-men-on-the-ice penalty, which they refer to as "Member over."

When showing up at the rink, it always seems to be a mystery to me as to whether it will be a practice, a friendly game – with just about any level of competition – or a league game, which are full contact and come complete with a referee, two linesmen, an announcer, statisticians, and even goal judges. After smiles, bows, a few friendly greetings, "Komba-wa" (good evening), followed by watching for which sweaters the other players pull out of their bags, I eventually figure out what type of skate it will be.

As for practices, why adult hockey players, especially old guys like myself, even bother practicing is beyond me. Once on the ice, lining up on the blue line before a game, bowing and banging our sticks on the ice as a salute to the other team seems reasonable enough. But bowing to an empty player's bench after a game always feels really weird to me. It's nice to be several inches taller than the other guys, particularly in the full contact games, but my size doesn't seem to make me any better of a target for receiving my teammates' passes. It seems that my Japanese teammates get nervous when they see me standing wide open in front of the net – I suppose that it is a little intimidating for them to see my big, lily-white face, round blue eyes eagerly waiting for a pass.

There are a few notable differences in the arenas in Japan compared to what I am used to in Canada. A stop at the concession stand is a real eye-opener, featuring non-traditional hockey rink delicacies like barbecued squid-on-a-stick, barbecued or raw octopus (Detroit Red Wings fans would love that), and rice balls wrapped in seaweed. The rinks all have Zambonis, but it is also common for players to grab brooms made from a bundle of twigs tied to a bamboo stick to sweep around the corners after every game or practice. It is also common to see a lot of players with backpack-style hockey bags. These come in very handy for the many guys who ride motorcycles or scooters to the rink, or have to carry their equipment through a myriad of trains and subways.

To bring some sanity into the wacky hockey experience in Japan, I was lucky to find the Tokyo Canadians Hockey Club. Formed in 1995, it turns out that there were several other Canadians living and working in Tokyo who were also getting frustrated playing on

Japanese teams. I joined in 2001, the year I went back to using normal hockey terminology like "nice save" and "nice shot".

Team membership has grown over the last few years and the Tokyo Canadians currently have a 35 player roster, which includes four Japanese and three Americans. Besides playing hockey, many team members get together for ski trips, golf outings and nights out on the town. Several of the team's favourite watering holes are now active team sponsors. The Tokyo Canadians are also active participants in charity events and raise money for a variety of charitable organizations.

Top competition for the Canadians is hard to find in Tokyo. The best contests are actually the Red-White games, where the team divides in half to scrimmage amongst each other. Playing the Japanese university teams is always a laugh. The youngsters typically arrive at the rink two hours before the game, accompanied by their groupie-looking girlfriends dressed in long matching parkas with team colours and logos. For at least an hour they stretch, jog, and do calisthenics. During pre-game warm-ups, the young university players whiz around the ice yelling and chanting. Noisy as can be. However, once the game starts, the out-of-shape Tokyo Canadians quickly take control of the game and show that seasoned playmaking skills are too much for the high-spirited youths. Time and time again, the Tokyo Canadians teach the Japanese university teams a thing or two about playing the good old game of hockey.

In addition to playing exhibition games in Japan, the Tokyo Canadians also participate in international tournaments throughout Asia each year. This includes tournaments in several tropical destinations like Malaysia, Singapore, and Thailand – places where most people can't even imagine having sheets of ice to play around on. The road trips for tournaments are a real blast. The hockey is competitive, and you get to see and experience parts of the world where you would not expect to end up with a hockey bag over your shoulder.

I'll never forget the puzzled looks on tourists' faces at the Chiang Mai airport in Thailand when they saw dozens of guys unloading hockey sticks and hockey bags off of the baggage carousel. Seeing bags and sticks strapped to the top of a tuk-tuk (a pick-up truck converted into a taxi) was also an unusual sight. Our first trip to Chiang Mai was during the Loi Krathong Festival (Festival of Lights), which is an experience in itself. During the festival, beautiful candles float down the waterways, giant lanterns fly through the sky like UFOs, and fireworks can come at you from just about any direction. Crossing one small bridge in town felt like something out of the movie "Apocalypse Now". Midway across the bridge, while we all ducked and dodged firecrackers that were being thrown at our heads and feet, one of our guys somehow managed to get himself run over by a moped. No injuries, just lots of laughter that brought tears to our eyes.

Chiang Mai was good to the Tokyo Canadians, as we were champions of the 2002 Thailand Ice Hockey Tournament. However, that victory turned out to be a turning point. Since that tourney, the Tokyo Canadians battled their way into the final of the next six tournaments that they entered, only to return to Tokyo with a second place trophy. One year before the final game in Chiang Mai, against a strong team from Switzerland, one of our overly pumped up players said, "Just in case someone needs to know, I've had six Sudafed and about four Red Bulls." That energy helped the Tokyo Canadians to score quickly against the Swiss. Although we tired rapidly, and a certain player become freakishly

The Tokyo Canadians
Back Row (left to right): Cam Knox, John Richmond, Rob Smaal, Brent Carlson, David Lotocki, Joji Hiratsuka, Stu Kimoto, James Heather, Denton Venable

Front Row (left to right):
Dave Lindsay, Scott McCaskie, Hiroki Narushima, Jason Young, Gary Cox, Mike Larose, Koichi Ogiso, Satoshi Chawanya

www.tokyocanadians.com

pale, we held on to win the championship.

No such luck for the hung-over Tokyo Canadians in the final against the Seoul Geckos Glaciers in the Korea tournament of July 2003. Or against a talented Finnish team, composed mostly of Finns who work for Nokia and live in Singapore, in the next Thai tourney at Bangkok in November 2003. Nor against our arch rivals, the Hong Kong Selects in Singapore in February 2004. Then against a young University of Minnesota alumni team in Las Vegas in April 2004, our squad (average age of 37) was winning 3-2 early in the 3rd period but collapsed in the closing minutes. The cagey Geckos once again clipped us in Seoul in July 2004, and later that year we watched the star-studded and unstoppable Siberian All-Stars skate circles around us in Bangkok.

The winning drought came to an end for one half of the Canadians, when Team White defeated Team Red in the championship game for the Tokyo Canadians Hockey Club Samurai Sword in December 2004. The other half of the squad is still thriving for another tournament championship.

We all really look forward to participating in tournaments. Both for the hockey and for the whole road trip adventure. Three Tokyo Canadians even travelled to Ulaan Baatar, Mongolia, to play in a late-January tourney on an outdoor rink in frigid minus 40 Celsius weather. An old bus, with a poor excuse for a heater, not only provided transportation to and from the rink but also served as dressing room, player's bench, and penalty box. Instead of a Zamboni, or yack-drawn ice making equipment, a wet rag on the end of a broken hockey stick was used to groom the bumpy, uneven ice surface. It sounds like most meals consisted of a "Mongolian warrior plate" which consisted of meat, meat, and more meat. The guys returned with a great selection of fur hats fit for Genghis Khan. After the trip, they all said that the weather was brutal, the hockey was crappy, but they were all really glad that they had made the trip.

The Tokyo Canadians now make regular annual appearances in the Singapore tourney in February, the Korea tourney in July, and the Thai event in November. Plus we send a team, or at least a few players, to a few other tournaments throughout the year.

At our first visit to the Korea tournament, the host team, the Seoul Geckos Glaciers, perfectly executed their strategy. They bought our players drinks all night long, and had their two female players talk guys into participating in numerous drinking challenges. As a result of the long nights of partying, two of the Tokyo Canadians didn't wake up in time to make the final, and the rest of the team definitely lacked energy. One lesson learned while in Seoul, in addition to the need to drink in moderation, was that although there are red and white striped poles all over town, it isn't easy to get a haircut. It turns out the poles indicate a massage parlor, not a barbershop.

Before moving to Japan, I spent four years living in the Middle East and played hockey once a year in a tournament in Dubai. It's a great tournament, with some really fun parties each night. At the 2002 event, four players from the host team, the Dubai Mighty Camels, planned to perform at one of the parties as the Mighty Camel Toe Jam Band. After the leading scorer for the Dubai Mighty Camels demonstrated that he couldn't play the drums well enough to even get a song started, friends and teammates booed him off the stage. I jumped in behind the drums and ended up playing about 15 songs with the band. It was a really fun time for me, even with my limited teenage garage-band drumming skills. Then at the 2003 Dubai tourney, I got to fill in on the drums again after the band's drummer got stuck saving lives and cleaning up after a disaster at the Dubai dry docks where he worked. That was the same year that a CBC television crew was there, so I even got to play the drums on Canadian TV – for about two seconds in a documentary called "The Hockey Nomad."

The Tokyo Canadians entered a team in the 2005 Dubai tourney. I didn't attempt to embarrass myself on the drums this time, but I was glad that my buddies from Tokyo were able to experience the excitement of the tournament. The Dubai contest has some unique features, like the legendary "Salmon Run" on banquet night. A folding wall opens up at midnight and hundreds of Russian and Chinese working girls from the adjoining disco battle their way through the crowd towards the banquet room full of nearly 200 drunken hockey players.

The tournament at Kuala Lumpur, Malaysia, was an interesting trip, especially for our hockey sticks. When we arrived at the airport, our sticks were nowhere to be found. Trying to explain what hockey sticks looked like to the airport staff didn't go too well. Finally, they found our bundle of lumber lying on the tarmac. At the airport, in a land where not too many people even know what hockey is, we were surprised to see advertisements for the tournament claiming that 20,000 people would attend the games, and more than 200,000 were expected to view the final. Sure, a lot of people showed up and watched from the four levels of the shopping mall that overlooked the rink, but I am certain that the numbers didn't come close to adding up to the promoter's claims. Back at the airport after the tournament, which was shortly after the terrorist attacks of 9/11, our hockey sticks once again got lots of attention. All of our sticks were X-rayed, visually inspected and then X-rayed again. Even after all of this, the inspectors still seemed extremely suspicious, perhaps because of the wicked, illegal banana hooks on my sticks. Then, upon arriving back in Japan, customs agents asked me if my hockey sticks were golf clubs. I answered "Hai" (yes), and was quickly on my way.

In addition to all of the fun playing for the Tokyo Canadians, you never know who might show up at the rinks throughout Asia. Gordie Howe was on hand at the Hong Kong tourney in 1998; Neal Broten, an ex-NHLer and a member of the U.S. Olympic team that became known as the "Miracle on Ice", played in the 2001 Thai tourney at Chiang Mai; a CBC crew followed the Rheostatics guitarist and recreation hockey fanatic Dave Bodini around the Dubai tourney in 2003 for the "The Hockey Nomad" documentary; Russian ex-pros entered the 2004 Thailand event; NHL Hall of Famer Mike Bossy came out to coach the U.S. Navy team in a friendly match against the Tokyo Canadians in January 2004; and Canada's GlobalTV showed up to film the Tokyo Canadians for a segment on its national broadcast, featuring one of our players who made his way from Moose Jaw, Saskatchewan, to Japan to become a kindergarten teacher.

I've had lots of good times with the Tokyo Canadians, and I look forward to more great hockey and more fantastic trips.

An Interlude… in Hoi An

Pam McKay

Hoi An. Quintessential Vietnam. Hoi An was previously known as one of South East Asia's most important international trading posts and is now a UNESCO World Heritage Site. The narrow streets are lined with fine art galleries, restaurants offering Vietnamese cuisine, small museums and expert tailors. It is all a joy to discover, so as a solo traveller with a couple of free days in Hoi An, I ventured out to indulge my passions for shopping and meeting the local people.

On a bustling morning, I hit the market and explored some of Hoi An's fabulous custom tailoring shops. As I had the time, and wanted to have several suits made, I did the rounds, and selected three shops that would share my business. I ended up with a nifty little schedule: place an order, visit a museum; place another order, visit a temple; get measured, go for a coffee and baguette... quite the whirlwind.

When it came to fittings, one shop hit it right-on – a raw silk, two piece suit, fully lined, and it fit like a glove – all for $32USD. Another shop had to make a few adjustments, and the last shop had to re-work a couple of things on my garments. A woman, in her thirties ran that particular shop, and her daughter probably about thirteen assisted her. They took the measurements and had fabrics on the premises, but the actual sewing was done elsewhere. A man on a bicycle (probably the woman's husband) would arrive, take the garment, and pedal off for 20 or 30 minutes each time an adjustment was made. While waiting, I spent a couple of hours with the shop owner and her daughter drinking tea and talking about life in our very different countries.

Did I mention the shopkeeper did not speak English? No matter – her daughter translated, and we had quite a chat. They found it strange that I was travelling alone for five weeks and my husband didn't mind. It was almost inconceivable to them that not only did I not have any children, but I did not wish to have any, and did not intend to. With a look of utter disbelief, the woman said quietly, "All Vietnamese want children." She spoke a little English after all.

It was a lovely afternoon, and I treasure the memory. As for my suits, the workmanship was good, and I still wear them after many, many years. A totally satisfying experience, I would return to the vibrant kaleidoscope of people and landscape in Vietnam any day.

Fast-forward… I can't believe that trip to Hoi An occurred over seven years ago. With my extensive passion and knowledge of Asia, I have become an Asia specialist for a major tour company, allowing me travels to the Far East such as China, Hong Kong, Japan, Korea; the deep pockets of the Southeast such as Indonesia, Thailand, Cambodia, Myanmar, Laos and Malaysia and South Asia's incredible India and Sri Lanka. Here are a few of my favourites I can share with fellow Canadians travelling to Southeast Asia.

Best food: Thailand
Most Impressive Monuments: Cambodia, Temples of Angkor
Top Towns for Wandering: Luang Prabang (Laos), Hoi An (Vietnam)
Lost in Time: Myanmar
Most Exciting Modern Structure: Petronas Towers Skybridge, Kuala Lumpur (Malaysia)
Most Tranquil Temples: Bali, Indonesia
Best Shopping: I can't decide – everywhere has amazing local products and cheap!

Overall, I have found Vietnam to be a beautiful, gentle, place. The people were warm and friendly. Many times, hearing I was Canadian, they started to speak to me in French – especially the older people. There is a real energy in Ho Chi Minh City, and even out in the Mekong Delta, things kind of 'bustle', especially at the floating markets. Hanoi was a little calmer, but still busy, and very interesting. Halong Bay was simply spectacular. My next trip will take me into the northern hills and overland into Yunnan, southern China, sure to be another incredible journey!

Pam is Tour Division Manager with Conference Tours

The World is a book
and those who do not travel
read only a page.

Saint Augustine

Susan Smith's Southeast Asia Sojourn

Susan Smith

A three-week trip to Southeast Asia took my husband Keith and I from the sublime – a luxurious resort in Thailand – to the ridiculous – an elephant ride in Cambodia. Along the way there were many wonders to behold. Let's start with Singapore bling. This 646 square-kilometre island off the tip of the Malay Peninsula can best be described as one giant shopping mall. The British after all, colonized it as a trading outpost in the 19th century, and since then it has definitely been trading – up and down and in all directions. The skyscrapers soar, while a labyrinth of underground shops creates a parallel universe beneath the city.

...Entry: Singapore Bling... Today I exhaust myself shopping and am forced to collapse by the pool, blistered and light of wallet. While I realize that this may not elicit much sympathy, let me assure you that shopping in Singapore is not for the faint of foot. I feel like a medieval pilgrim from the provinces who just made it to Chartres cathedral. I have shopped the magnificent Chicago; Bloomingdale's in New York and the Ginza in Tokyo-but there has never been anything like this. Picture The Eaton Centre on a massive amount of steroids. And, auspiciously, it is the month of what is called the Great Singapore sale. Being a bit cautious in new surroundings, I start out in the Raffles City mall attached to our hotel. Here, the mall itself is worth writing home about. Designed by architect I.M.Pei, the famous Chinese-American architect as part of the sprawling Raffles City, it has an airy four-storey atrium with palm trees and a beautiful fountain built as an abstraction of a campanile or clock tower. There are lots of familiar names here like Brooks Brothers, Tommy Hilfiger, Polo and the like. But there are some new discoveries such as British India, a high-end shop with a distinct colonial flavour; Robinsons's, a British department store, and several shops with clothes by local designers.

Having tested the shopping waters, I grab a cab to Orchard Road, the main shopping strip. With one gleaming mall after the next to choose from, I ask to be let off at the Takashimaya Shopping Centre, a main temple for practitioners of shopping. This mammoth and also beautifully designed mall has a separate store for every designer label known to man. The most sophisticated display? Tiffany's, of course. An avowed Japan-ofile, and still aglow from a great haul of kimono-based accessories, I head for the inner sanctum, Takashimaya of Japan.

In housewares I find bed linens another particular weakness-that are the most beautiful I've ever seen, made from fabric that could only have been designed in Japan. I am nearly overtaken with consumer fervor until a mental vision of a Canadian customs officer forces a sober second thought. Now to the third floor-women's clothing-another glimpse of heaven. But, here I am truly brought back to earth. I soon realize that this store caters mainly to the Asian fashionista-a tiny creature-and there are a lot of Size two's on the rack. Since I cannot fit one thigh into a Size two, I sheepishly ask the clerk if there are any larger sizes. She smiles

Royal Palace in Bangkok

patronizingly, in a way that makes me feel only slightly humiliated, and comes out with a Size 16! Now, being a Canadian woman with what is euphemistically called a 'full figure,' but I am no size 16, even if it is cut small. Wounded, but with my head held high in defiance, I move on to jewellery, where I am relieved to find that the size of my neck is not egregiously inappropriate. I decide on a pale jade necklace-on sale of course-and I am happy.

Across the street at Lucky Plaza, so named I suspect because you might be lucky enough to afford them, another dazzling display of consumer mania. This place specializes in jewellery-diamonds, gold, coral and jade. Here, the farther back you go, the cheaper the prices and the more there is to choose from. I find myself standing agog at an intersection of three stores where strands of pearls in every colour imaginable hang on racks like scarves. This is too much. I stagger outside to get some air and rest a while on a bench. Composure regained, I grab another cab and head for Bugis Village, billed in the guidebook as the place for the bargain hunter. Here, stall after stall in a makeshift outdoor market offers $20 skirts marked down to $10, $10 watches marked down to $5, and $5 T-shirts marked down to $2.

A huge crowd is trying to squeeze its way through a maze of narrow laneways. Bugis Junction, as the neighbourhood is called, got its name from a warrior tribe of Malaysians who immigrated more than a century ago to the island. In the 1960s it was known as the transsexual and transvestite area. Now it is the area of really cheap stuff. Just when I think there's nothing to buy here, I spy a beaded necklace in a shop of Nepalese imports. The price tag says $30, but the keeper of the shop says I can have a special deal of $20. I offer $15 and the necklace is mine. It's a full-spectrum day of shopping in Singapore. After Bugis, the only thing left to do is sit by the pool and have a Singapore-brewed Tiger beer.

...Entry... The King and Us... Today, we arrive at the airport in Bangkok to the news of more bombings, this time much closer than the carnage on the London subway. In the southern Thai province of Songkhla, a bomb has exploded killing another 50 people in a campaign of terror that has claimed almost 1,000 lives in the past year. Muslim extremists are suspected, as they have been waging war in an attempt to separate from the predominantly Buddhist Thailand. Security is tight at the airport, but it still does not rival Singapore during a visit by the International Olympic Committee.

After clearing immigration, we are relieved to be met by a small Thai man waving a sign with our names printed in large letters. Paul has been sent to meet us by Circle of Asia; the tour company that helped us plan this trip. More than a tour guide, Paul is a handler-he is

used to leading large groups on jungle treks-and he approaches our care and feeding with great efficiency.

After a brief but cacophonous airport experience, he ushers us into an air-conditioned car driven by Montri, a soft-spoken Bangkok native who has been driving tourists around some of the most crowded streets in the world for 28 years. We were in good hands. As we settle in for a three-hour ride to Hua Hin, a resort town on the Gulf of Thailand where we plan to grab some serious beach time, Paul explains that we need not worry because all the bomb attacks have been in a Muslim-dominated area to the south. Actually, we are heading south and will be spending the next three days in one province north of Songkhla. But these are not thoughts we want to dwell on. To further ease our apprehension, Paul points out that King Bhumibol Adulyadej, or Rama IX, is living in Hua Hin and that the Thai people would not be housing their much-loved, 77 year old monarch in a place that wasn't safe. We ask about the king and are given a thorough lesson in Thailand's parliamentary democracy and how it compares to Britain's. We pass the first of what will be several royal palaces seen on this trip. This one belonged to Rama IV, the king's great-grandfather, the one immortalized in The King and I. I reflect that when Anna Leonowens came to what was then Siam in 1862 to teach English to the royal wives and children, this country was a vastly different place – closed off from the rest of the world and suspicious of foreigners. These days it's what's known in globalization-speak as a NIC or newly industrialized country, hungry for foreign investment and eager to cater to tourists. King Bhumibol was actually born in the United States when his father was studying at Harvard.

Paul explains that Thailand is now a modern country with freedoms not shared by some of its neighbours in Southeast Asia. Yet not so far under the surface, there's a tradition-loving society in which the King still has veto power over laws passed by Parliament. Because of the unrest to the south, he is being asked to approve anti-terrorist measures that will limit new-found civil liberties. And while there is certain amount of press freedom, it remains illegal to report on the personal lives of the Royal family – the Windsors should be so lucky. As for The King and I, the video of the musical is banned in Thailand because it is considered disrespectful to the King. As we pass by the many billboard – sized portraits of Rama IX that keep appearing on the side of the highway, I can't help but think that his great-grandfather-caught up in a power struggle with an independent, reform-minded Western woman-would perhaps consider this his last laugh.

We arrive at the Dusit Polo Resort; we find all the talk about the King has been somehow prescient. It's a place fit for a king and we are suddenly being treated like royalty by an ultra-attentive and pampering staff. It is rare that a hotel is actually better than it looks in the brochures, but this one has exceeded our wildest dreams. Except for the destroyer that can be seen floating in the Gulf near the horizon in the vicinity of the real King, reminders of a troubled world are melting away fast.

...Entry... Buddha's of Unusual Size... Bangkok was founded in 1782 by the first King Rama, who needed to move his capital south from the central plains of Thailand to escape from the encroaching Burmese. People think its name means 'City of Angels', but that came later. The name Bangkok actually derives from 'bang makok,' which means 'Village of the wild plum.' Today at 600 square miles and more than 10 million people, it would more aptly be named 'City of wild motorbikes.' The diesel-choked streets grow more clogged each day as Thais migrate to the commercial hub from villages and smaller towns. The main mode of transportation is no longer the barges that drifted along a system of canals, or 'klongs,' that gave the early city its other nickname, 'The Venice of Asia.'

While some of the canals have survived, they are no longer used as transportation arteries, and sadly, the last floating market has disappeared from inside the city.

Instead, two-wheel Honda and Suzukis weave through the streets with great agility, squeezing themselves in between cars and trucks and into other unlikely places, three to a lane.

On our way out of the city after spending two nights on the 19th floor of a hotel surveying all this action from a great height, we are fortunate to be reunited with our guide and driver, who have a nice big air-conditioned car. As it's nearly 40 degrees outside, we count our blessings. It takes more than an hour for the congestion to thin out but eventually the sprawl gives way to smaller roads that lead north, for a trip back in time-not only in terms of living standards, but also in terms of cultural richness.

While Bangkok is an amazing city, the true spirit of Thailand still seems to be rooted in the countryside, where Buddhism still has a firm hold and people seem as concerned with accumulating merit for the next life as they are with accumulating wealth for this one. Each little village we pass, even the very poor ones, manages to support a temple, with the distinctive Thai decoration on the roof, which symbolizes either flames, or the naga snake of Indian mythology, or the horns of a bull, depending on who you ask. And most homes proudly display a little temple mounted on a stick, rather like a birdhouse, except these sometimes elaborate creations are believed to house the spirit of ancestors and deities standing guard over a particular location. Co-existing with the more formal practice of Buddhism, they are places of daily offerings of flowers, food and incense, all gifts to the spirits.

Today many people are going to the temples to pray and make offerings to the Buddha because it is a national holiday marking the beginning of a three-month period for the monks. But these pilgrimages are not necessarily solemn occasions. At one point we're taking a break along the road and are caught up in a small, festive parade complete with floats and a contest to see which artistic offering of flowers and candles will be considered the best in the area. An estimated half of all Thai men take monastic vows for a period of several weeks to several years before they marry and enter the workforce and support for the monks of the community is strong. More than 90 percent of Thailand's population consider themselves Theravadan. While this branch of Buddhism is sometimes called the 'little vehicle' there is nothing little about the role it plays in everyday life here. Representations of the Buddha, in fact, are so large and so plentiful in this part of the world that no number of pictures and travel guides can prepare the visitor for their magnitude.

After a morning's drive we make it to the ruins of Ayutthaya, an old capital founded in 1350 that was presided over by 33 rulers until it was abandoned in the late 18th century after a defeat to the Burmese army. At one time, the city supported a million people, whose inhabitants traced their roots from all over the world. The renovated ruins, now a protected world heritage site, are a fascinating look into the beginnings of what became the classical Thai style with tall chedis, or stupas, used to house relics and ashes of senior monks and other important people of the time. The giant conical spires have a haunting quality as they speak of a time of greatness that ended in disaster. Outside what were once the city gates, new temples are filled with life, especially today. As we remove our shoes to join lotus-and incense-bearing pilgrims entering one such place of worship, we are stopped short. We expected to see a Buddha, of course, but what we actually see is a hand-a hand so massive that it takes a while to catch on to what we are seeing. When we do look up, way up, we see that it belongs to a giant Buddha wrapped in his own giant saffron robe silently accepting the offerings with his beatific, inscrutable, Buddha-like expression. The presence of such a big Buddha stops the mind, which I suppose is exactly what it was intended to do.

Back out in the light, tour guide Paul, himself a devout Buddhist who wears a reliquary around his neck on a heavy chain, explains that the Thai people seem so cheerful because, while they may not be wealthy, they have enough to eat, and they have the Buddha. It's an interesting theory and in this context I tend to believe he might be right. As we continue to make our way to the north, the countryside starts to sprout rice paddies and fields of soybeans and other crops and I get my first glimpse of a live water buffalo, who isn't too keen to be sharing his road with a tourist.

Next on the agenda is the town of Phitsanulok and the Wat (or Temple) Mahathat, with another giant Buddha, this one made of bronze and standing. Here there are long rows of smaller gilded Buddhas and we are treated to some monks in their distinctive saffron ringing the gong that signals a period of practice. It is a lovely moment that captures the spirit of the place and of the day. But not being monastically inclined, our materialistic minds turn to thoughts of dinner and a nice soft bed.

We've covered a lot of ground today and the heat and humidity are taking their toll. Paul, however, has other plans for us and insists that we can't yet call it a day.

We've seen the Buddhas sitting and standing, and we must see one reclining, one who also happens to be one of the largest in Thailand. Paul is not an easy man to argue with, especially when it comes to Buddhas. Historians aren't sure how this Buddha got here or what once surrounded him, but he clearly rules over this part of the countryside with great equanimity. Paul explains that the Buddha in this position is one who is ready to enter nirvana, or a state of enlightenment, known as complete spiritual wakefulness. But my husband and I mischievously think he's wishing two tired travellers to have a nice long sleep.

By Susan Smith, Writer/Editor in Toronto.
(First appeared in www.globeandmail.com)

Running with the Bulls in Sumatra, Indonesia

Graydon Hazenberg

As soon as I read about them, I knew I had to go see the bullfights! Despite numerous trips to Europe I had managed not to see the feria in Pamplona, so I decided that now was my chance to live the life of Hemingway.

The bullfights near Bukittingi, however, in the hill country on the west side of Sumatra, Indonesia, are a bit different from what you see at a Spanish corrida.

For one thing, it's not a man-against-beast affair. Two water buffalo bulls lock horns with each other, and no humans are involved in the actual combat. Hemingway would probably not have approved. Another difference is that the fights take place in a large open field surrounded by a flimsy fence, with the spectators lounging on the grass either just outside or, for the brave, just inside the fence.

It was four o'clock on a Tuesday afternoon, an hour before starting time, and the crowd was gathering for the weekly bullfights in Koto Baru. Having arrived an hour earlier on a local minibus from Bukittingi, the local backpacker's hangout, I watched the crowd slowly assemble.

I had been concerned that the fight might be a contrived spectacle, strictly for the gawking tourists, but the 500 or so people milling around were almost exclusively Indonesians. Male Indonesians, to be precise; I couldn't see a single local woman. The men were a mixed group, that is, old-timers in their black cylindrical Suharto hats, 40 something in shabby clothes and floppy hats, well-dressed 30 something and cool 20 year olds making the scene with elegant hairdos, fake Ray Bans and shirts unbuttoned halfway to the waist.

In the last hour, four bulls had been led into the ring and allowed to wallow in one of the mud holes scattered around the grassy circle. Most of the crowd clustered in knots around the combatants, probably sizing them up for betting purposes. A lot of cash was being flashed about by the local bookies, and everyone seemed to be betting. Meanwhile, tinny loudspeakers blared Indonesian pop songs that were so distorted that they sounded like the Muslim call to prayer. In the distance, far from this cacophony, the verdant slopes of Gunung Singgalang, the local volcano, rose against a clear blue sky.

As the starting hour neared, the crowd swelled to somewhere around 2,000. The tourists maybe 30 in total, stuck out tremendously, their blond heads towering over the sea of short Indonesians. We gravitated towards each other to chat and compare impressions.

Westin, an American who was staying at the same guesthouse as I, kept trying to get a surreptitious photo of a man in the crowd around one of the bulls. The man looked like an Indonesian Don King, complete with finger-in-the-light-socket hairdo. Although, our imitator required a small mesh hairnet to reach the heights of the real Mr. King's coiffure. It was unclear if this Don King was the promoter of the bullfights.

Finally, without much fanfare, the first bout began. The owners led their animals through the circle of onlookers, into the middle of the ring. The bulls immediately lowered their heads, almost to the ground, and locked horns to a tremendous roar from the crowd. A tight circle quickly gathered around the two bulls.

My guidebook had said that the bulls didn't get seriously injured in these fights, but that's not what it looked like in the first fight. The bulls frequently pulled away from their clinch and twisted their heads to gore their opponents with the tips of their horns. As the fight progressed, cuts opened up on the necks and cheeks of both bulls. When they butted heads, the sound of clashing horn and skull was enough to make me flinch! Both bulls soon had blood oozing from the tops of their heads between their horns.

The fights continue until one bull gives up and runs away. After the initial few minutes of sparring and locking horns, the bulls seemed to tire, and spent most of their time eyeing each other while the owners and the referee shouted encouragement. In contrast, the pace of betting in the crowd quickened as the bout wore on.

I scrambled around the outside of the crowd, trying to get some clear photos of the bulls. Twice I skipped away in a slight panic (along with all the people standing near me) when it looked as though one bull was about to make a run for it. This bull, the bloodier of the two, started to turn his head away from his opponent, almost as though he were offering his neck to him: "Go ahead. Gore me! See if I care!" His eyes rolled back in his head, giving him a crazed look, and then without warning, he swung his head back toward the other bull, their heads meeting with a sickening crash.

Slowly the other bull moved Eye-Roller back into one corner of the ring, where earth ramparts six meters high enclosed the bulls and allowed hundreds of people to look straight down onto the fight. I climbed out of the ring to get a better view from above, but just as I was about to take a picture, Eye-Roller decided that he had had enough punishment. He turned his head and fled at top speed, hotly pursued by the victor.

They raced through a muddy watering hole and out the gate in the fence, disappearing out of sight into the village. Dozens of people ran behind, trying to catch the runaway animals. Westin told me afterwards that when they bolted, the bulls headed straight towards him and he ran for dear life for a few seconds, before throwing himself to the side. Hemingway would have approved of this echo of Pamplona.

The second bout was a different affair. The bulls were bigger, and a lot keener on fighting. They locked horns instantly and stayed like that for nearly the entire fight, which lasted for maybe three minutes, rather than the ten that the lightweights had taken. Like two sumo wrestlers, they tried to force each other straight backwards by sheer strength, their legs straining and their hooves straining for grip in the muddy parts of the ring. Their neck muscles had to be astonishingly strong to endure the twisting that they were inflicting on each other while they locked horns. I gave up trying to take pictures, as people were clustered so closely around the bulls that I couldn't get a clear shot. I stood and watched, waiting for another spectacular fleeing finish.

This time, the fight ended with little warning. The loser suddenly turned tail and fled, but instead of heading straight out of the gate, he began to run laps around the inside of the fence, before crashing through the gate and continuing the orbital maneuvers around the outside of the fence. Onlookers scattered to either side of the onrushing mass; I wondered idly how many people get run over every year by fleeing bulls.

Then it was all over, except for the paying of betting losses, which provoked at least one punch-up between a loser and his much larger creditor. We tourists congratulated ourselves on having come to a genuine spectacle of a vibrant local culture, and piled into bemos for the ride back to town.

Westin was grinning from ear to ear after his narrow escape, and we relived the moment over cold Bintangs back in Bukittingi. It wasn't quite a bottle of Rioja in a Pamplona bar, but it had a touch of Hemingway anyway.

Graydon Hazenberg of Thunder Bay, Ontario is currently travelling around the world.

My aim is to put down on paper,
what I see & what I feel
in the best & simplest way.

Ernest Hemingway

Indonesia's Apple-Less Garden of Eden

Margaret H. Dickenson

As our helicopter skims over a carpet of jungle trees, excitement rushes through every ounce of my body. Our official visit to Kalimantan would include visiting a colony of orangutans and a jungle search for the extremely rare black orchid.

Momentarily, my appetite for this incredible adventure is dampened as drops of warm blood from a sacrificial chicken are sprinkled over our bare feet. This is part of the local welcoming ceremony, being performed by a nearly naked, hauntingly thin and sinewy medicine man. I soon see that it is a reflection of the unique cultural richness and ethnic diversity of a nation spread across some 15,000 islands. My husband Larry's assignment as Canadian ambassador to Indonesia would offer many such exotic experiences.

With my self-taught, "functioning" level of Bahasa Indonesia (a deliberately chosen unifying national language), I mixed with locals across many of the 27 provinces, each with its own traditional costumes, textiles, food, dance, architecture, customs and ceremonies.

On the island of Bali, there was an endless round of religious and private ceremonies and feasts. Contrasting that culture were tourist-filled beaches and luxury hotels, thriving communities of artists and some of the most romantic resorts in the world.

Life in Indonesia was fascinating. With high temperatures all year round and only two seasons, the vegetation was extraordinary. So were the floods. Within days of our arrival, the disadvantage of a patio pool became obvious as we frantically dammed water lapping at our patio doors. Still, we appreciated dining beside the pool in the still of the night, watching nature perform her own version of Cirque du Soleil: Salamanders scrambled up and down walls and across ceilings, bats dive-bombed the swimming pool as they feasted on water bugs. More intimidating was the procession of "civet cats" that slunk along the top of the tall cement garden wall after having preyed upon their domestic neighborhood cousins.

Indonesian food offered its own seductions. I detected Arabic, Chinese, Indian, Dutch, and Portuguese influences and the spices (nutmeg, cloves, cinnamon, pepper, cumin) upon which those countries made their since-lost fortunes. Indonesian cuisine has complex flavours and includes hot chili sauce, fresh ginger, garlic, lemongrass, coriander, peanuts, palm sugar, coconut, coconut milk and rice. Barbecued and fried foods are popular, as are glutinous desserts prepared in a myriad of colours, shapes and flavours. Street carts offer everything from homemade snacks and drinks of all kinds to full meals.

Like Eve, I lusted after the infamous Durian. Its reputation of tasting like heaven and smelling like hell is well known. Public buildings and some airlines have signs banning its entry while Larry did the same on our property in Jakarta. However, on a trip to North Sulawesi, a driver put one in the trunk for me while we attended a banquet. My plan was to taste it later in the garden of our inn. But as soon as we got into the car, Larry picked up the outrageous odor. I knew I was in trouble.

The flesh did taste delicious, and the stench persisted until the end of our stay. But, the aroma was not just Durian. The clouds, which never left the peaks of the mountains, were caused by gas escaping from volcanic activity. The smell of rotten eggs permeated everything. On our last night, our hosts roasted an entire pig for 12 hours in an underground pit. So delicious, juicy and tender was it, Larry forgot the Durian episode.

First appeared in Diplomat Magazine. Dickenson is an International Award Winning Cookbook Author & TV Presenter living in Ottawa.
www.margaretssenseofoccasion.com

No one realizes how beautiful it is
to travel until he comes home
& rests his head
on his old familiar pillow.

Lyn Yutang

Indonesian Moon

John S. Goulet

Mount Rinjani

Rinjani is an active volcano that built the island of Lombok along the Indonesian fault. I think that is a realistic way of looking at the role of a volcano. When anyone mentions Rinjani they always say it is on the island of Lombok. Wrong. The island is on the volcano and consequently on the tectonic plate fault line that has spawned the longest and most active line of volcanic reactions on earth. Everybody who lives on the island of Lombok owes their existence and their livelihood to the volcano. The inevitable if irregular eruptions not only created the land they stand on, but also enrich the soil for a bountiful harvest. With the tropical rains that fall on the island come the lush emerald green forests and terraced rice fields that symbolize this exotic region of Indonesia.

Ying and Yang, good and evil, life and death, and all the other opposites best describe the condition of living on Lombok or within the vicinity of any other volcano for that matter. Shaken by the sporadic and unpredictable eruptions, it's no wonder that the people of Lombok worship the spirit of the fiery inverted-cone shaped mountain. They celebrate its very existence and are careful not to enrage the spirits in charge. Don't rock the boat. Don't spit into the wind. And don't tug on Superman's cape.

I don't think I could live like that. Always tiptoeing around your creator. Always being a quiet good little boy as not to wake your step-dad. Always offering to light the alter candles. Forever throwing white rice, green banana leaves, and saffron coloured flowers into the abyss. Although I took out my Grandma's garbage on a regular basis, I was always the one to rise to the challenges of our neighborhood. For example, I led the raid on old man Apsit's garden. Not just because he had the best garden around, but mostly because he guarded it so jealousy. He basically told us we were not allowed to go there. Mistake.

What is so tempting about the 12,296 foot Rinjani volcano, however, is that somewhere in the distant past an enormous explosion tore off the top of the 15,000 foot volcano and left a gaping crater large enough to place a small city within. With thousands of years of tropical rain the vast crater basin now holds the large beautiful Segara Anak Lake whose surface sits at about 6,300 foot elevation above sea level. Both the Balinese and the Muslim Sasak of Lombok Island make a pilgrimage to toss ritual rice and goldfish into the clear mountain waters. At the risk of appearing irreverent I will say that this scared lake was begging for a bush pilot to pay homage.

Lake Segara Anak is about five kilometres long and three kilometres wide with a 200 year-old secondary cone protruding out of the east end. Adding to the expansive feeling of grandeur the surrounding crater rim extends upward another three to four thousand feet to create a continuous enclosure looking so much like a crumbled wall of an ancient Roman coliseum. The only exception is where the initial explosion, or later erosion, created a narrow gap in the north end. Here the excess rainwater flows out forming a waterfall and a river that flows to the sea.

The normal way of getting to this wondrous lake and for that matter to the volcano peak is to hike up. Time magazine did a feature on this particular climb in 2001 and declared it to be difficult, treacherous, and extremely worthwhile. The adventure is akin to climbing Mt. Kilimanjaro, but tougher. It takes two days of slogging up a tropical rainforest trail complete with heat, insects, thieving monkeys, and washed out footpaths. On the morning of the second night you awake early to make the rim at sunrise. Here you can view the lake at its most serene before the inevitable mountain top clouds form to cover the basin for the remainder of the day. If you want you can hike down nearly 3,000 feet to the lakeside campsite, but few hikers do. Even fewer actually make it to the 12,296 foot summit.

Despite being considered "worthwhile" it is also considered dangerous. 11 people have died making this climb in the last five years and ten have died in the last three years alone. You have to be in very good tropically acclimatized condition to even make the rim in relative safety. The ideal then of hiking down to the lake and back up again puts most hikers off, despite what they might have thought before they left base camp in the village below.

I had no intention of making the hike into this timeless abyss. Not when I can have 675 horses pull me up. In fact with 675 horses I could have pulled up Nero and his chariot as well. Yes, I consider myself the decadent explorer. With two bottles of water, a dozen morning fresh donuts from the Dunkin Donuts shop at the airport, and the Pilot Operating Handbook, I started off to conquer the "coliseum".

This was booked as a training flight with Alex, one of the Indonesian pilots that I was training on the amphib Caravan. When I told him to flight plan us for a Rinjani over-flight and when I jumped in the Captain's seat, he knew immediately what I was up to. To his credit, however, he did not ask. It was one of my early morning spur of the moment decisions that I did not wish to discuss with anyone least they attempt to talk me out of "it." "It" being whatever I am going to do that I am really not supposed to be doing. I hate busy bodies that place logic and reason in front of everything and spoil the magic of the moment by making me realize the futility or wickedness of the pending act. Especially if they are in a position of authority, for instance, my boss or the director of civil aviation.

Just in case you are thinking that I am totally lacking in respect for authority I must mention that I did have the "go ahead" to attempt a landing on the lake in Rinjani. The authorization went something like this: "You know John we have been thinking about this landing for some time. Of the last three pilots we hired as consultants, one said it was outright impossible, one said it was most likely possible, but foolish, and the third said it was possible, but he wouldn't try. We would sure like your opinion, as you would be the most experienced of any of them. When you get a chance see if you couldn't give it a go."

Now that sounds like a case of clear-cut authorization to me. So on this beautiful clear Balinese morning we took off for Rinjani. The Area Control cleared us up to our requested 11,000 feet and during the 30 minutes it took us to reach the volcano I leafed through the POH. The only relevant section I was concerned about was the settings for Take-off Power. I quickly found out that the charts, contrary to what one pilot had said, did go up to and beyond 7,000 feet. (Not that I was about to let a little extrapolation get in my way.) In fact, the charts claimed that we would be able to get full power, ie. 675 hp., on takeoff. Zounds!

As the Rinjani volcano was along our normal flight path from Bali to the luxury resort island of Amanwana, run by the Aman resort group, I had surveyed the lake many times before. So this morning I knew what I had to do. I approached the crater from west to east and flew past the northern side. That put the V shaped cleft where the overflow from the lake spills out as waterfalls on my right (starboard) side. The tops of the V were varied from 9,000 feet on the west to 10,000 feet high on the east climbing toward the 12,296 feet pinnacle, while the bottom of the V was at the lake level of approximately 63,000 feet. That V shaped "cut" made up my entrance.

I could have spiraled down from the top of the crater rim, but I was planning for the dramatic. I wanted the entrance into the crater to be a grand entrance and flying through the relative narrowness of the "cut" into the enormity of the "basin" would be like flying into another world and another time. "Ladies and Gentleman, this is your Captain speaking; we have just left the present and returned to the beginning of time."

I flew past the cut and the waterfalls at about 10,000 feet and did a slow left descending turn, away from the volcano and over the numerous terraced rice fields far below. The smallness of the villages and fields were to give a perspective on the enormity of the volcanic mountain we were about to enter. I descended to 8,000 feet and finished my 270 degree turn facing due south, straight into the V shaped entrance. From this vista we could see the lake nestled inside the cone. It looked small and I really had to fight the illusion that the crater basin would be too small to circle inside. Proximity claustrophobia began to set in. That is when you are flying between mountains and the canyon walls seem to be closing in when in fact they are not. If you panic you will attempt to turn when you should keep going straight ahead.

In the turn I had slowed the aircraft down to 120 knots and before I entered the crater I selected 10 degree of flaps and slowed down to 105 knots. Slowing down gives the passengers a better view and lowering the flaps gives me a better view over the nose. As we entered I banked and turned right to follow the rim just below the edge, giving the sensation of being inside the basin without getting too low too quickly. Once inside I prescribed a slow left turn passing the 12,296 peaks on my right and placing the 7,838 feet Gunung Barujari secondary daughter cone on my left. As we were still prescribing a left hand arc this allowed us to bank over the crater of the daughter cone and have a good look inside the vent. We could see the steam and smoke rising from the still active cone.

In fact, when we entered the main crater Alex startled me by hollering, "Look, the volcano is erupting." Just below the daughter cone was a large fresh flow of lava differentiated by the older flows by the blackness of the lava and by the fact there was no vegetation growing on it. It was indeed a new flow in volcano time, but that was eight years ago already in our time. Alex was convinced that it had happened since our last over-flight on the way to the resort the day before. I said that if it was recent the water would be boiling and steaming. "Oh, yeah," he admitted.

As we circled I had a good look at the water. My first concern was of course the depth. With the water sitting in an expansive cereal-bowl like hole, you don't have to be a geologist to extrapolate that the lake would be deep, possibly very deep… reportedly nearly 800 feet deep. If I landed anywhere away from the crater sides I would have deep water. Besides I could readily see to untold depths in the relatively clear water.

I admit, though, I was slightly concerned that the water might have been contaminated with sulfate or chloride, which are both corrosive and which can both be found in volcanic waters if the area is active. The water, however, looked like any mountain lake in the Canadian Rockies. The water was a slightly opaque blue green that appeared clear despite the obvious dissolved minerals obscuring the light transmission.

The real clue as to the safety of the water was the colour and health of the trees near the shore. Even from 2,000 feet I could see that the trees growing along the waters edge were a healthy green, and not yellow as they would be if the water had turned poisonous. Moreover the trees acted as our "canary" sentinels that if there had been any recent serious chemical activity then the vent would have been issuing carbon dioxide and killing the nearby trees. There were no dead trees.

As I passed between the peak on my right and the daughter cone on my left I set up for my final approach. Flying once around the inside of the rim was partly for effect, but it also served a useful purpose. It allowed me to gauge a feel for the winds that might have been blowing over and around the peak that would affect my final approach. This morning, and I suspect most mornings, it was dead calm. In fact, I would be contending with glassy calm. The fly around also allowed me time to gain perspective on distance and height.

I flew past the entrance, which was also to be our exit, and over the open water of the main body of the lake. I made a tight 45 degree bank to the left and swung around to the backside of the daughter cone again. The 180 degree turn served two purposes. For one, it allowed the passengers on the right side of the aircraft (that is when and if we ever flew passengers into the crater) to have a look inside the active cone and secondly allowed me to set up for a full flap descending final with a slow right directly onto the main body of the lake. Dropping down low over the backside of the active cone allowed us to spot the many steaming vents and hot springs that we had not seen from higher up.

It was only as I came onto final descent that I realized how big the crater and ensuing lake really was. I knew it was something like five kilometres long, but confined inside the caldera's rim it had looked small. In fact, without proper reference I would have said it was too small to take-off from. The problem was in attempting to find a perspective to judge actual distances and relative sizes. A cliff or rock can be large or small. A tree can be large or dwarfed. The water had no perspective value and the air was so severely clear at this time in the morning that I could not judge distances properly. In essence I had no reference. I did not even know the actual water level and with glassy water how would I know to round off before taking a nosedive? I can see why the other pilots had held off. Every survival instinct told me to abort and go around. So according to my plan I did.

My plan was to get down as close to the outside shore as possible and use the shoreline as a reference to my height above the water while reading the actual altitude. This worked fine except getting my reference was still difficult. With the assumption that the trees inside a 7,000 to 10,000 foot crater would be stunted and small, I kept thinking I was close to the shore when I was not close at all. It took some concentration and conscious effort to get myself to fly the aircraft closer and closer to the near vertical cliff face shoreline to be able to properly judge my height above the water. When I did get close I realized that the trees were a lot larger than I had imagined. No wonder my perspective was thrown so far off.

Once I worked out the water's altitude I turned out toward the main body of the lake, skimmed the water with my floats, and headed toward the cleft in the crater rim. The landing and takeoff area could not have been blessed any more perfect. The area I chose was: 1) the longest straight stretch on the lake; 2) into any wind that would come up later in the morning; 3) in the direction of the hiker's camping site if we ever wanted to go ashore; and finally 4) facing straight toward the exit for both the takeoff and for any necessary go around.

As this was my practice "go around" and also simulated takeoff (without actually committing myself) I waited until about half way down the lake and then applied full power. Alex was waiting to take the readings of ITT (turbine temperature) and rate of climb. From 6,300 feet we eventually climbed to 8,000 feet with less than 740 degrees Celsius on the ITT (within the green) and over 500 feet per minute climb. Both well within acceptable limits. At 500 feet above the lake level I pulled the power slightly and headed out the exit to visualize what the escape would feel like. I could have stayed inside the basin and spiral climbed my way out the top, but flying through the narrow walls of the cut was again a way to feel the transition with a dramatic flare.

Exiting through the cut we flew right over the hiker's camping site, the hot springs, and the falls, and then the landscape dropped away. You could feel the rush as the earth left you suspended. We climbed just enough to make an "operations normal" report to the Area Control frequency and then I banked left and headed back inside. I did another approach and this time followed up with a full stop landing. Landing in that clear mountain water was like landing on a mirror. Struggling with spatial disorientation I felt like we were still suspended in air as I could see the reflections of trees and rocks and clouds below me.

After falling off the step I feathered the prop and Alex got out on the float to take a water sample. The water was as clear and pure as from any spring. I later teased Alex that he could quit flying for a living and start bottling and selling that water. "Alex's Rinjani Spring Water."

The feeling of having been the first person ever to land on this lake and possibly the first person to ever have even been on the water, as there was no evidence of a boat, did not sink in while we floated on the beautiful lake. All I could think of was getting airborne again. As soon as Alex was strapped in I applied full power and started my takeoff run. Although I was getting full power on the torque I was not getting full thrust from the bit of the prop. The thin air was certainly affecting both thrust and lift. I had to use every tiny bit of my floatplane experience to coax her onto the step.

Before applying full power, I pulled the control column right back with full up elevator deflection. The nose rose agonizingly slowly until it raised no more. At this point I relaxed the pressure on the control column allowing the nose to drop slightly. This helped the prop to "pull" the aircraft onto the step changing the center of buoyancy from the rear of the floats toward the center and allowing the airflow to build up over the wings creating lift. As the buoyancy moved forward I could feel the force of the water starting to push the aircraft onto the step and so I rocked the control slightly ahead to "help" her fulcrum her way into the planning attitude.

As soon as she gained the step I pulled back gently on the control column again to search for the "sweet spot" or the ideal planing altitude where neither the bows nor the keels are digging or dragging and the aircraft is riding solely on the smallest section of the V shaped keel. When she was ready to fly, I broke the right and almost simultaneously the left float from the lake's surface tension and was airborne. The rest was easy. Alex and I had earned our donuts that day.

Oddly enough we savored our conquest and did not tell anyone for a couple of days. When I finally ventured to mention to the head office in an "Oh, by the way" conversation, they were ecstatic. I was congratulated with a official fax from the head office thanking me for my effort. "For Travira Air, your landing inside the Rinjani volcano is equivalent to Indonesia landing a man on the moon." They could hardly wait to see the photographs. Alex and I looked at each other. We had no photographs. Neither of us had brought a camera. As there were no witnesses how were we to prove that we had actually landed on the lake? We did not exist. On a later conversation, the Operations Manager lamented us not having photographs. He smoothed the way for a second flight when he asked, "Do you think there is any way that you could get some pictures?" I said, "Sure, let me work it out."

So two days later Alex and I were off again. This time on landing I shut down the turbine and was left with an eerie silence as we ghosted to a stop. Alex and I got out to document our landing for prosterity. Alex had come up with the idea that we should shoot each other holding newspapers. He had even brought the original papers from the morning we had done the first landing, as that would be more authentic. Now we could prove that we had actually been there that morning. The irony was that the papers we held carried stories in Indonesian about Bin Ladin and the Al Quida shortly after the 9/11 attach on America. We looked like hostages being held for ransom.

The morning was so beautiful and calm that this time I did not want to leave. The opaque emerald green water was cool, and the air was crisp and refreshing. The trees turned out to be tall straight pine, looking like something out of "Myst" or descendants from the Mesozoic Period, which blended in well with the alpine atmosphere. The feeling was that of a summer morning in the Rocky Mountains. Alex said, "I've never been to Canada, but I imagine that this is what it is like." I said, "Yes… except for the volcano" towering over the east end of the lake. Silhouetted against the morning sun, I could make out the rising steam from the volcano cone. The steam appeared thin and wispy from the distance, but it was venting nonetheless. In other words, here was evidence that the volcano was still active and threatening. It was hard to believe that something so majestic and beautiful could be so potentially dangerous. In the calm cool of the morning, the tension I felt was not caused by the awareness of imminent death. The feeling was rather one of unfulfilled opportunity, like that of not having said goodbye to a dying father. I would be disappointed if Rinjani erupted one day and I was not there to see it go. I certainly understand how a volcano could come to be worshipped.

Because this flight was also booked as a training flight, I let Alex take over left seat so he could do a series of landings and takeoffs. I had had my fun. On the second takeoff we were flying out of the cut toward home when Alex looked down toward the campsite and said, "I think someone is mad as us, they are waving with both arms." I had learned many years ago that a two-arm wave means someone is in trouble. Much to Alex's amazement I decided to investigate. His reasoning was that if we were in trouble we should avoid the person. My reasoning was that if we were in trouble we should confront and reason with the person, but more likely we would be providing assistance to someone who needs our help. Alex made his first glassy water landing and we step taxied toward the shore. When he decided we were getting close to the shore he pulled her off the step, but to our amusement the waving person was a tiny little dot on the shoreline. We were still a kilometre away. The clear air and magnitude of the mountains had fooled our senses again.

We approached the shore and I had Alex shut down the turbine early. We pulled out the aluminum paddles and slowly worked our way toward the rocky shoreline. The bank was gradual making it easy to nose in the aircraft. With no waves or tide to worry about we only had to hitch a rope and leave her floating. The waver turned out to be Jaron Starling from Australia.

Jaron was a backpacker who had hired some local guides from the village below and set out alone to find this legendary lake. He said it took him three days to get this far when one of his guides came down with pneumonia. The poor guy had been holed up in his tent for several days with no sign of improvement. Jaron said the guy had a fever and could hardly breath. He certainly did not want to risk forcing him to hike three to four hours back up the crater rim and then another eight hours back down to the village. Jaron was getting low on supplies and he was concerned about his guide's health. Could we possibly carry this sick man back down to civilization?

In the first place, Jaron was our perfect reason for coming here. We were flying over and saw that someone was in trouble. That was the truth. Plain and simple. We had no choice but to land. In fact, that is what I had suggested to Travira Air when they wanted to know how we were going to get official permission to start a tourist run into the lake. Sell it as a safety backup for mountain climbers and hikers who might get in trouble while they were on top. If we were to drop in a couple of times a week it might even increase the amount of hikers willing to risk the journey knowing that they could be rescued if necessary. Medical evacuations and exploration started bush flying and here we could combine the two.

I agreed to take the man as long as he could be ready by 9:30 a.m. It was already 9:10 a.m. and the midmorning clouds were starting to fill the lower valley and spill out through the cut into the lake. By 10:00 a.m. the crater basin would be filled with clouds and we would be trapped. Jaron had the guide packed and ready to go by my deadline. We loaded him on board and Alex conducted a takeoff and spiral climb back toward the exit. Once through the cut we started a descent, slowly as not to affect our congested passenger, toward Mataram the capital of Lombok where we dropped him off.

By the time we got to sea level he could already breath easier and the colour came back to his face. After disembarking he shook our hands and turned to leave, no doubt dazed at the experience he had just been through. It took him three days to get up the mountain and less than 20 minutes to get back down. Janaldi Anar Farultuk Terhnir was on his way back to his family.

Watching him trod slowly homeward I felt like I had accomplished something more important than just being the first person to have landed on the Indonesian moon. Thanks to Jaron Miles Starling of Australia and his two-armed waved, we had no doubt saved a man's life.

John is the editor of Virtual Horizons
www.bushpilot.com

My Dinner with Trudeau

Margaret H. Dickenson

During our years in the Canadian Foreign Service, Larry and I hosted hundreds of dinner parties. We held one particularly memorable party in 1995 when former Prime Minister Pierre Elliott Trudeau paid a private visit to Indonesia. Larry arranged a three-week program for him and his party. With excursions to several far-flung islands, it would have exhausted a man half his age.

Mr. Trudeau was a remarkably animated and gracious guest. He insisted Larry call him Pierre. When Larry admitted that he would find that difficult, Mr. Trudeau, with that characteristic twinkle in his eyes, immediately declared that he would then always have to address Larry as "Mr. Ambassador". And, so he did. I went for the Pierre salutation.

Word spread quickly of Trudeau's visit to Jakarta, and invitations to our "welcome to Indonesia" buffet dinner became hot tickets. The guest list was drawn from Indonesia's political, business and cultural circles and included several interesting Canadians living in Indonesia.

Larry and I were eager to see what Mr. Trudeau would wear to dinner. Earlier that day, upon his arrival at Jakarta's airport, he was dressed casually – T-shirt, slacks, and bare feet in sandals. And for dinner, he wore the same sandals but was otherwise perfectly attired, wearing a lightweight suit with his signature rose in the lapel, shirt and tie. And this time, he added socks to the sandal look.

Known for his uncanny eye for the fairer sex, our guest radiated enthusiasm and appreciation as some of Indonesia's most stunning and intelligent women joined him at his table. During the evening, many guests waited their turn to exchange a few words with our famous guest.

At the end of a lively and most successful evening, the chair of the Canada-Indonesia Business Club discretely requested Larry's permission to give Mr. Trudeau a gift. It was a framed copy of a photo which had appeared in a Jakarta newspaper about 20 years earlier, during an official visit he had made as prime minister. It showed a younger Mr. Trudeau mischievously pulling a rickshaw through a busy street with a shocked and bemused rickshaw driver sitting proudly in the passenger's seat. Mr. Trudeau was touched by the gesture.

As guests bid good night, one of the ladies at Mr. Trudeau's table asked him about his itinerary. Upon mentioning that he would conclude his stay at the luxurious Canadian Four Seasons Resort in Bali, she promptly proposed hosting a traditional Balinese dinner at her Oceanside estate – a short drive from his hotel. He accepted only upon the condition that Larry and I would join him. We obliged.

What a romantic and magical evening that was. Torches glowed, frogs croaked, crickets chirped; gamelan music filled the still night and stars, the clear sky. As we feasted our way through several outdoor pergolas, guests marvelled at the parade of delicious local foods.

Huge green leaves, served as platters upon which the food was directly placed. The moon emulated that of paradise – serene, intimate, and mystical. To add to the exotic nature of that occasion, several Hollywood types (shooting a movie on a neighbouring estate) joined us. Mr. Trudeau held court as he told of his odyssey to the remote "Spice Islands" and Irian Jaya (home of the former head-hunters).

In the end, Mr. Trudeau was thrilled by these two bookend dinners.

First appeared in Diplomat Magazine. Dickenson is an International Award Winning Cookbook Author & TV Presenter in , Ottawa, Canada

Canuck Downunder

Valerie Close

I have been living here now for 21 years and am originally from Toronto but lived in the Yukon and Vancouver prior to moving to Australia. When I was growing up in Canada, I remember being in the library of Brown School in Toronto and reading about Australian summers and how hot it could be on Christmas Day. I remember reading a book that described Australians hosing down their homes on Christmas to try to cool them inside, as it was 40 degrees out. As a Canadian experiencing cold, snowy Christmases this was unimaginable.

I also remember reading Gerald Durrell and his adventures with animals from New Zealand and Australia. As a child the exotic animals of hot countries always fascinated me like the duck-billed platypus of Australia. When Colonial biologists first saw this animal back in the early 1800s and sent a specimen back to Britain, it was thought to be a hoax! So strange was this mammal with a duck-like bill and webbed feet.

When I first arrived in Australia I was staying in a suburb of Sydney called Bondi, an area famous for its beach. I used to go out for morning walks and the first morning I went out I heard this monkey calling from a backyard. I couldn't actually see it, I could only hear it. I thought it was interesting that someone had a monkey for a pet, as there are no monkeys native to Australia.

After that morning, each day that I went out for my morning walk, I would hear more monkeys. I was beginning to think it was quite strange how so many Australians seemed to have pet monkeys.

During this time, I was invited to an Aussie BBQ and I asked the locals about all the pet monkeys I had been hearing. The Australians looked at me like I was from another planet and patiently explained that they weren't monkeys at all but an Australian bird called a kookaburra. They found it very amusing and thought I was crazy to mix the two up!

After that night, I found out that I actually wasn't all that crazy, as the early Tarzan movies with Johnny Weissmuller used the taped calls of Australian kookaburras as monkeys calling in the jungle!

The ironic thing about this story is that for the past 14 years I have been an environmental educator teaching Australians about Australian flora, fauna and natural environments. After my monkey-like introduction, I quickly fell in love with the natural world here! Today I live in Sydney and my home backs on to bush land with the morning and evening calls of kookaburras, rainbow lorikeets, magpies and king parrots. We have our resident lyrebirds who can mimic all sorts of other Aussie birds as well as car alarms, chainsaws and lawn mowers! At night ringtail and brushtail possums visit the garden as well as bandicoots and swamp wallabies. It is an animal-lovers paradise and it all started with the sound of a "monkey!"

The Twelve Apostles

What I know about Australia

Locations… try to get out of the cities into the country or bush as we call it here. Far north Queensland, Central Australia where Ayres Rock and the Olgas is, land that is completely different to Canada. Sydney is worth seeing as it has the Opera House, a beautiful harbour which is bordered by National parks and wonderful beaches. For any Canadians who are into wine, the Hunter Valley, which is about 1.5 hours from Sydney, has a number of vineyards with wine tasting of quality Australian wines. It's a beautiful area with gently rolling hills and gum trees.

Another location that not many tourists would go to is Lord Howe Island down south. Most of the islands that tourists know and visit are around the Great Barrier Reef. Lord Howe is not tropical but it is beautiful and laid back.

Food… the suburbs of Balmain, Glebe and Crows Nest have a number of restaurants that the average tourist would not know of, with a wonderful selection of restaurants to choose from. One of the many things that amazed me about Sydney when I first moved here was the number of restaurants and how they could all do enough business to stay open!

Food to try… Sydney Rock Oysters, lamb, Barramundi fish and all our fruit is fabulous especially the stone fruits in the summer (December to February) and mango. We also have some great Aussie cheese from a number of areas such as King Island.

Indigenous people… there are a lot in Sydney in most suburbs. However, one of the main residential areas for Aborigines is a suburb called Redfern, which is near the central area of Sydney. Indigenous people maintain and manage the land that surrounds Ayres Rock and the Olgas in Central Australia. Their interpretation of the meaning of the land is fascinating and gives anyone visiting an insight into over 40,000 years of Aboriginal knowledge of the land.

A few other funny tips… if a Canadian is socializing with Aussies are; if an Aussie asks you to bring a plate to a BBQ or party, they don't mean just bring an empty plate, they mean bring a plate with some food on it that you can contribute to the meal! If an Australian tells you to "grab your cosies, thongs and shout them a stubby". It means grab your swimsuit, your thongs and buy them a beer on the way!

A few colloquialisms… that Aussies use and a Canadian visitor may hear; "struth" meaning amazing, "fair dinkum" meaning honestly, no kidding, "flat out like a lizard drinking" meaning you were very busy, "dry as a dead dingo's donger" meaning you're very thirsty! "You right?" means can I help you. "Too right" means that's for sure.

Australians call the bathroom, the "toilet" (that's not considered crude here) and if you are in an older style Australian house you'll find the toilets are separate from the bathroom area so if you ask to use the bathroom, you'll get there to find no toilet! It will be in the next room! That always threw me when I first arrived. I eventually learned to ask for the "toilet."

All Australian males call each other "mate" because they can't remember each other's names! Australian males love to bond around the BBQ. Australian males and females have a great sense of humour. Australian parties will have the males chatting together at one end and the females at another. It doesn't matter what age, socio-economic level or that it is now the 21st century. It still drives me batty – tell me it doesn't happen in Canada!!!

Australians think Canadians sound Irish, even the Irish think Canadians sound Irish over here! Or they'll call you American - get used to it! Until you say or they ask you to say "out and about". That will clear you of any American connection!

Valerie Close is an environmental educator living in Sydney, Australia.

There are only two ways
to live your life.
One is as though nothing
is a miracle.
The other is as though everything
is a miracle.

Albert Einstein

South Pacific Adventures

Bert Ollivier

My beautiful wife, Juanita (a.k.a. my "B.W.") and I embarked mid November of 2005 from San Diego, California aboard a luxurious cruise ship on the first leg of a four and a half month adventure. Our carefully planned trip involved a combination of sea, air and ground travel which would include Mexico, South America, South Pacific Polynesian Islands and on to Australia, culminating in the North and South Islands of New Zealand.

Prior to leaving home, good friends warned in vain that we would be totally bored after only a week or so on a cruise ship which for 35 days was the first part of our journey. We did not listen and thankfully those well-meaning travel advisors were wrong.

How could any passenger be bored when we had weekly lifeboat drills, a real fire alarm after 10 days at sea, plus a code yellow situation followed by a code red strict warning issued by the Captain at the 20 day mark? But more on these emergencies later in this story.

Boredom was also headed off at the pass by activities and events happening hourly and occasionally on the half hour with such frequency that my B.W. and I had to split up in order to take in certain activities that appealed to us most. However, I hasten to add at this juncture that nothing was forced on us and no tiresome and annoying P.A. announcements were made to interrupt our naps or visits to the library as has been the case on other cruises we have taken.

All activities were listed in the daily printed program to read at our leisure and for us to decide upon. The choice of activities was an interesting mix of sports, an exploration series, port talks, and items of history such as how the Inca Indians were exploited in South America including the Spanish influence. These were just a few from the cruise director and her staff's bag of tricks. The entertainment staff hailed from England, Canada, the United States, Australia and New Zealand, all under the kind and watchful eye of the vivacious and talented cruise director, an elegant lady from Newcastle on Tyne, England.

The ports of call were all interesting mini-adventures interspersed thoughtfully between the many days at sea. First calls were in the Mexican ports of Cabo San Lucas, Zihuatanejo, and Acapulco; then on to South America, visiting Manta, Ecuador, followed by Salaverry and Callao/Lima in Peru and on to Chile's Easter Island (Rapa Nui). Pitcairn Island followed where descendants of the infamous mutineers on the H.M.S. Bounty boarded our ship. Next, were visits to Tahiti, Bora Bora, Pago Pago in American Samoa, and finally New Zealand where we stopped first at Paihia in the Bay of Islands followed by Auckland, our final stop. A little more detail and summary of the highlights of the ports of call will follow later in this story.

As our ship docked for two nights in Peru many passengers took advantage of a three day, two-night trip to Machu Picchu, high in the Andes. I declined the opportunity having visited this most interesting, ancient Inca city many years ago but after interviewing several who took the trip I regretted not doing a re-visit. One of the main reasons for regretting my non-move is that the method of getting there has vastly improved by the addition of three pampered hours on one of the great train rides in the world and I am a train buff. Passengers returning from the tour told me that the journey calls to mind the luxurious standards of rail travel in days of yore featuring opulence and elegance. Two dining cars and a bar car assisted in making the Hiram Bingham train a truly unique experience. The train is named for the explorer who "found" Machu Picchu about a hundred years ago. The bus ride from the train station to the city on a switchback road was a bit of an adventure in itself according to most passengers. This part would not have appealed to my B.W. I know from experience!

The construction technique in Machu Picchu, utilizing huge blocks of stone all fitting together perfectly, free of cement, is a wonder in itself with none of the technology used in building today. Spanish conquistadors came to this area in search of gold but they never found the ancient, Lost City of the Incas.

Two nights in Cuzco were spent at a first class hotel, one before the Machu Picchu ascent and one after. A comprehensive tour of this ancient city was part of the deal set up by the ship's excursion desk. Some of the passengers who felt that these, and other tours, were over priced re-named the tour office the "extortion desk". Personally, I felt that was going a bit too far and that the excursion desk dispensed good, honest value, with well organized tours of the many ports of call.

The visit to Peru now brings me to one of the emergencies I alluded to earlier. The ship's officers and cruise staff had suggested that we not partake of any food ashore, especially at "off the beaten track" restaurants, bars or cafes. It followed that folks who took the planned excursion to Machu Picchu and Cuzco would be exposed to food ashore but at hotels and eating establishments chosen by the cruise company. However, because the ship was docked for two nights in Peru, many passengers went ashore on their own. Coincidentally, four days after leaving Peru, a few cases of GI (Gastro Intestinal disorder) occurred prompting the captain to issue a Code Yellow situation. This is a cautionary procedure meaning that the captain warned all passengers to wash hands more frequently and avoid shaking hands with anyone.

The code yellow was okay for a short time but then several more cases of GI escalated the problem to Code Red, a much more serious problem. This resulted in practices such as salt and pepper shakers being replaced by individual tiny paper packets, cutlery distribution by staff only with latex gloves on their hands, along with no touching of food at the buffet service areas, and so on. Any person diagnosed with GI was confined to his or her stateroom but with unlimited room service for food, DVD's for entertainment and anything else the person required or desired within reason. Quarantine usually lasted no more than 48 to 72 hours, so it wasn't that bad. Sanitation hand cleaning devices are always installed at the entrances to all dining areas as standard fixtures. However, the captain ordered all staff to make sure passengers used them. Also, the captain asked all to sing "happy birthday" while washing, the result being thoroughly washed and rinsed hands.

It turned out that the captain knew exactly what he was doing because within a week, he happily declared, much to the relief of all aboard that the epidemic was under control and the Code Red was cancelled.

There are two emergencies that are serious aboard a cruise ship, the first being fire and

Waitman Harbour, Auckland

the second a Code Red alert. We experienced both. The Code Red as detailed above was interesting but not life threatening.

But now for a more exciting part, the fire alarm. Actually this happened before the Code Red, on the tenth day out of San Diego. The alarm sounded at 11:30 p.m. and my B.W. bolted out of bed, dressed immediately and donned her life jacket, all in a matter of about a minute and a half. To this day, she argues with me about my version of the events, but that's my story and I'm sticking to it! Her version is that she was quite calm, dressed slowly and at the same time cajoled me for not taking action similar to hers. Regardless, I decided that because of the life jacket drill when the captain went over the rules that state: "women and children first," I went calmly back to bed and waited for further announcements. Five minutes later the captain announced that the alarm was caused by smoke in the laundry room and that the situation was completely under control. He then apologized for awakening us and reminded us again that the situation was completely under control. In retrospect, I must admit that my overly calm reaction was probably not correct and I made the appropriate apologies to my B.W.

I should emphasize at this juncture that readers should not be wary of taking a cruise because of the emergencies described above. Conversely, the captain went through an exercise in statistics to prove that being a passenger on a cruise ship you are many times safer than staying at home and certainly much safer than using other means of transportation, especially the family automobile!

The only other unusual side activity involved a medical emergency, which took place off Tonga shortly before midnight on the 18th day. It is not unusual on a cruise ship that caters to older folks that a medical emergency is serious enough that a passenger is sent ashore for hospital treatment. As a journalist I tried to find out the cause of the problem but was told very politely and firmly that the information was private and personal. Of course I understood. I had no choice. However, my B.W. and I watched the tender being lowered with all searchlights on. The doctor and a nurse accompanied the patient with two ship's officers and three sailors to handle the embarkation and landing of the tender. We were told that a helicopter was waiting for transfer to a hospital in Tahiti. After the tender had left we were enthralled by the number of fish that had gathered by the ship's hull, no doubt attracted by the powerful lights. They looked so thick it appeared a person could walk on the water with the support of the fish and I wondered if there is any connection to that familiar story in the bible? All this interest was heightened when my B.W. brought my attention to the moon, which suddenly appeared at the stern as an abnormally huge orange ball. This phenomenon apparently occurs frequently when close to the equator.

Now for a summary of port highlights other than Machu Picchu, which I have already, described in detail. Heading an unofficial popularity contest among the passengers was the visit to Pitcairn Island. I took it upon myself to conduct a survey and I was only a little bit surprised to learn that Pitcairn took the lead especially in view of the fact that no landing took place nor did we anchor. There is no suitable dock for anything larger than a longboat on this forbiddingly rugged island and no facility for tendering. Instead, the captain idled the engines and we hovered fairly close to shore from our arrival at 9:00 a.m. until departure at 12:30 a.m. A longboat containing 37 Islanders out of the total population of 50 boarded our ship at 9:30 a.m. They were loaded down with baskets, backpacks and oversized valises stuffed full of everything from T shirts to small carvings of the Bounty to walking canes made and autographed personally by descendants of some of the Bounty mutineers. Even postage stamps at face value were for sale. Our crew thoughtfully arranged several long tables for the islanders to set up shop with their wares. Their tables were barely set up when pandemonium set in, almost unbelievably, as the passengers feverishly bought up almost all the islanders' wares. The frenzy was similar to a Boxing Day sale at home. The calm approach of the Pitcairners, who were patient, trusting and honest to a fault, was a refreshing difference from the frenzy created by frantic customers. Hundreds, perhaps thousands of pictures were taken that day of the sale activities plus the approach and departure from Pitcairn Island.

Other Polynesian ports of call were Bora Bora, and Tahiti but the next most popular after the Pitcairn experience was Pago Pago. Spectacular scenery made this stop a good one and from a modern history standpoint many passengers recalled a movie starring Rita Hayworth as the fictional proprietress/madam of the infamous Sadie Thompson's Inn. The Inn is now a small boutique hotel, bar and dining establishment. Many passengers made a point of visiting the bar for a bit of libation and nostalgia.

All passengers enjoyed seeing and hearing folkloric entertainers who boarded the ship at all ports from Mexico through to New Zealand.

A personal "high" enjoyed by my B.W. and me was the appearance of a certain tugboat upon our landing at the Port of Auckland, New Zealand. We knew that one of our nephews was a tugboat captain and that he was to meet us and drive us to the airport for the next part of our journey. However, we did not know that he would go so far as to meet us in his tug. What a surprise when we saw him through our binoculars! We waved from the promenade deck and he acknowledged us with a short sounding of the boat's horn. But that's not all. When we disembarked he said he could not drive us to the airport as he had another docking job to do right away. We told him we had lots of time to spare before our flight so he said we could come with him to see first-hand how tugs work and then he would drive us to the airport later. He knew I was an experienced boater having owned eight in my life and suggested I could drive the tug into the open water of the harbour. My B.W. was apprehensive and nervous, especially when encouraged by the mate to put one engine in reverse and the other in forward and performed what is known as a "donut" … what a thrill! Upon approaching the freight container ship it was suggested that my nephew and his mate take over, much to the relief of my B.W. We were told that the job is easy if the weather is good but it can be a nerve wracking experience when the waves and wind are high combined with driving rain. I decided right there and then that any future boating I do will only be of the fair weather variety. I sold my last boat five years ago so even that is unlikely unless we win a lotto.

This concludes the first part of our four and a half month South Pacific Adventure.

Bert Ollivier is a certified travel counsellor (c.t.c.) in Canada and the U.S.A. He was President of one of Canada's largest travel agency and wholesale tour operation. Most recently he was Managing Director, Canada for the State of Hawaii Visitor & Convention Bureau and is now a published travel journalist. His home is in Nanaimo, British Columbia.

A Canadians' Escape – Kiwi Style

Pam & daughter Hannah Layden

Kia Ora! It is a strangely common phenomenon that lures us away from this Great North Country of Canada to far away, remote, and sometimes very familiar places. Prepared to meet the odds or unexpected we delight in finding that common ground and friendships that promise future visits. Those we choose to accompany us hopefully enrich travelling this route. In my plans I chose to take my thirteen-year-old daughter, Hannah.

Like many, I too long for "greener pastures" and New Zealand is all that and more. New Zealand "Land of the long white cloud, steaming volcanoes, lush rainforest, rolling green pastures and snow-capped mountains." As a parent I was bound to the promises of "wouldn't you rather see New Zealand" long before Hannah's friends would inevitably speak of Disney or my own friends shared news of Disney "Deals." As a realist, I felt experiencing bathing in collected rainwater and seeing people actually wearing their rubber boots would hopefully enrich her soul. As well, I had urgency for Hannah to meet friends that Murray, my husband and I made years ago between these two countries.

The timing was based on "airline sales" and "school's out" thanks to a probable teacher's strike in Red Deer, Alberta. We decided on four fairly loosely scheduled weeks and two small bags. Our agenda, to visit friends, tour the country, relax, and hopefully get to Milford Sound this time as that was one location we did not see when we were in New Zealand over 20 years ago. Milford Road has the most magnificent mountain scenery leading to Milford Sound where you can cruise the majestic fiord. The famous Mitre Peak, natural scenery, dolphins and seals in their natural habitat are all along the road. The popular Lord of the Rings trilogy contributed to the popularity of the South Island with its spectacular Southern Alps, dramatic glaciers, rushing river gorges and deep winding fiords.

As in the past we couldn't wait to see the Bell family of New Zealand; we met Joy Bell many years ago when she came to Red Deer to visit Anne, her sister, both in North America as part of the Rotary Club exchange students program. We arrived in Auckland, the 'City of Sails' on the north island, to be greeted by Joy, and her two children, Alex and Rose. Their home is a gem situated just outside of Warkworth, north of Auckland, on a stretch of island just thirty miles across by horse or car. Of course, bunking out in the loft of an old school house, amongst antique collectables laying around for twenty years or more, and greeted by pet ducks, the Jack Russell Terrier, "Jack" and a pet lamb named Ruby, gives New Zealand that "rural appeal". To complete the paddocks are a couple of ponies, two horses, one of which is a true wild horse rescued from the great Carmanah Peninsula roundup, named "Sally."

Our evenings occurred after the children were in bed, excluding Hannah of course. These twilight gatherings often included, yes, wine and lamb, always wine. For the gourmet connoisseur New Zealand is a haven for food and wine lovers. Between 1994 and 2002 the area under grapes in New Zealand has increased from 6,680 to 18,247 hectares and continues to rise as winemaking supersedes the sheep industry. With New Zealand's "clean, green" image, grape growers and winemakers aim to keep it that way by protecting the environmental integrity of their wine production with sustainable wine making and vine yards. With wine, of course came many stories of Sally's escape from the roundup and of Buster's escape to the neighbour's, which through tragedy lead to a very fine meal of lamb. And of course there were Mr. Bell's many stories of Africa. As a young man he worked in Africa prior to meeting Joy's mother and immigrating to New Zealand. New Zealand offered him great fishing, sailing, and retirement living next door to Joy where he shares paddock and clothesline.

It was at the dinner table that I experienced the graciousness of my daughter's presence. Amongst the adult chatter, her face beamed in the candlelight, taking it all in. And as dishes were being cleared, her enthusiasm extended as many stories of Africa continued. Joy's neighbour also shared New Zealand history. It was his family that formed the community just over the hill. In his 60s, this well respected sheep farmer told nightly stories of sheep rescues, dog handling and dog shows, and managing farms, some even for status seeking Asians. We felt privileged to know this man of genuine warmth and practicality, like our New Zealand family. And we were grateful for everything shared.

Joy, a native New Zealander, while in California as a Rotary Club exchange student, rode up to Red Deer, Alberta, on a rather small Brand X and very yellow motorbike. In New Zealand this would be the smart way to travel. (I preferred the bus until I was overcome with switchback sickness.) I was grateful for my backseat driver shouting LEFT! Two lanes, only, also helped keep this Canadian on the right track. In Auckland, Joy had owned an antique store and surrounded herself with the most interesting collectables that would make most minimalists cringe; even her toilet was a masterpiece displayed in resin. As an artist, she has succeeded in the arts and crafts circle. This is not what most Canadians would consider to be arts and crafts. As we searched for some cultural enlightenment, we were directed to the craft shops. Thankfully, this meant watercolours and the like and not papier-mâché. The trend at the time was mosaics, which we found in simple aluminum teapots covered in colourful bits of porcelain, and with 'gallery' prices. Recently, I received a clipping of Joy's success, a life size mosaic couch that has drawn great interest. There seems to be an opportunity for New Zealanders to capture the market of everyone looking for something different and unique.

It is the ruggedness of New Zealand that calls for the simple life yet many felt compelled to capitalize on modernizing. Unlike 20 years prior, Californians offer lattes to surprised Canadians like myself. There is a striking difference between the original and the modern New Zealand yet it offers a comfortable, sort of shabby chic feel. For the New Zealander, this could bring conflict. Having found a piece of countryside to call her own, our friend found herself up against one of the country's newest industries, a tree farmer from the north island city of Wellington. By herself she was able to put off the destruction of her piece of paradise as well as becoming a true heroine in her community. Since the land is considered useless after a limited number of harvests there is concern for those who wish to continue to view this beautiful countryside It was a lesson in courage and we had a glimpse of how tough New Zealand women could be. And as far as I know, Joy is still trying to break Sally.

Our travels concentrated on coastal roads. Often the only roads that provided magnificent views were atop hills and cliffs! Even in the Korangahake Gorge (site of the original gold rush of 1875) we needed to arrive at the very top to scrutinize which direction to descend upon. Of course daylight is a must. I had travelled these roads once before and this time I tried to imagine what this must feel like for Hannah. I had hoped there would be enough adventure without the bungee jumping and black water rafting! I would have been very content to swim with the dolphins had the ocean cooperated. Along the way, since pubs were off limits to 13 year olds, I was fortunate enough to sample some of New Zealand's dark beers, not found in Red Deer stores.

The hostels welcomed all ages though they preferred guests in their early 20s. And while Hannah may have been surprised and perhaps flattered to be invited to the local pub one night, I was astonished at my quick response which must have sounded "parental" and brought quick apologies from the girls. If that didn't remind me of my years gone by, being handed the task of fire chief, complete with keys and discount lodging did.

I had hoped that the generation gap between mother and daughter would be narrowed on this journey and we both feel completely rejoiced in the memories created. With New Zealand, Hannah and I shared openness for the unexpected, a willingness to share all responsibilities and decisions. We made new discoveries together sharing in the exhilarating beauty of the Island. Even sharing a bed let alone a bedroom never brought a raised eyebrow. And most of all, never in 30 days did the words, "clean your room," pass these lips.

The success of this journey continues as letters are written, plans made to return with friends and most of all, Hannah cultivating a love of travel. After this visit to the other side of the world, our dreams to travel to Europe, take university in Australia or even New Zealand is very enticing and not so remote or unattainable. New travel memories and stories will always be created and written, but this first trip will always be special and exclusive to Hannah and myself! We never did make it to Milford Sound… hmm another journey to the edge of the Pacific.

Pam resides in Red Deer, Alberta with her husband Murray and their three children, Hannah, Jacob and Quaid.

Mermaid Pools

Scott and Lizbeth

Mermaid pools at Tutukka Northland, New Zealand Island of New Zealand, an absolute paradise to visit. We spent six glorious years in the thermal; reminded us of Prince Edward Island, our home!

New Zealand Pavlova

Provided by Louisa Powell

To many New Zealanders the Pavlova is a classic Kiwi icon.

Recipe
3 egg whites, 3 tablespoons cold water
1 cup caster sugar, 1 teaspoon vinegar
1 teaspoon vanilla essence, 3 teaspoons cornflour

Preheat oven to 150°C. Using electric mixer beat egg whites until stiff. Add water and beat again. Add sugar very gradually while still beating. Slow beater and add vinegar, vanilla and cornflour. Line an oven tray with baking paper. Draw a 22 centimetre circle on the baking paper. Spread the pavlova to within 2 centimetres of the edge of the circle, keeping the shape as round and even as possible. Smooth top surface over. Bake Pavlova for 45 minutes, then leave to cool in the oven. Carefully lift Pavlova on to a serving plate. Decorate with whipped cream and fresh fruit.

Louisa Powell is a Kiwi Canadian living in Calgary, Alberta.

Trek to the Antarctica

Tom Grehels

I am standing on top of a windblown cliff. Below me the enormous caldera of Deception Island stretches from one copper coloured cliff to the next. On the black volcanic beach penguins and seals compete for a place in the sun. Overhead a skua patrols the skies while a blue iceberg lazily floats by in the open ocean. This is Antarctica, the seventh continent and the last frontier of adventure travel. I make my way down to the beach and strip down to my swimming trunks to take a dip in a hole we dug in the sand, which by now has filled up with warm water from the hot springs. As I sip on a drink, cooled with ancient glacial ice scooped from the sea, I think back to the adventures that have come my way over the past few days.

I travelled to Antarctica on the Akademik Ioffe, also known on as the Peregrine Mariner. Originally built in Finland only a few years ago as a Russian research vessel, this comfortable ship has been chartered by one of the world's foremost adventure travel companies, Australian based Peregrine. Staffed with a friendly Russian crew, knowledgeable and enthusiastic Canadian, Australian and British guides and fitted with modern comforts such as a sauna, gym, bar and pool, this ship was to be my home for the next nine days.

We left Ushuaia at six o'clock in the evening with the sun still high in the sky. One of the strangest phenomena of sailing the Southern Ocean during the summer is the amount of daylight you get. The sun comes up long before anyone is awake and doesn't go down until eleven in the evening. At night it never really gets dark and often travellers can be seen on deck in the middle of the night watching the scenery and wildlife go by. The amount of light gives you enormous energy and sleep is often the last thing on your mind.

Soon our small group of 32 travellers and 40 crew left the calm waters of the Beagle Channel behind and entered the stormy seas of the Drake Passage. Days are livened up by sightings of many different species of Albatross, Petrels and whales, as well as by the lectures of the competent guides. Their enthusiasm is unmatched in the industry and their love for the continent is evident all throughout the trip. On board we had a naturalist/ornithologist, a historian, a photographer and an expert on tourism management, each of whom gave informal lectures and slide shows in their own field of expertise.

The most colourful person on the cruise was probably the Russian captain. Stern and with a permanent look of bemusement about the antics of us landlubbers on his grizzled face, he was the king of the bridge! I ran afoul of him several times as he caught me whistling aboard his ship. Being a superstitious sailor he would come up to me and poke a weathered finger in my face and proclaim loudly "Njet!" It was only much later that I found out that whistling is thought to bring on severe storms.

After two days on the open seas, we entered the sheltered waters around the South Shetland Islands and Antarctic Peninsula. On board we had some German scientists

Tom on Deception Island

whom we dropped off at Bellingshausen, a Russian research station. From here we sailed on towards our first landing site at Mikkelsen Harbour. Here a deserted Argentine research station sits on a low hill on an islet surrounded by majestic mountains. Chinstrap penguins have made their home all around and they don't seem to be the least disturbed by visitors. I made my way between two colonies to the other side of the islet where a Weddell seal was lazing in the sun. It turned onto its side curiously observing me as I sat down on a rock taking in all the scenery around me. Enormous mountains capped by blue glaciers formed the backdrop to a picturesque bay littered with floating bergie bits of all different sizes and shades of blue and white. A Chinstrap penguin emerged only metres away on a slippery rock, lost its balance and, fell over backwards. Shocked, this clown-like creature dove back into the water. Back on the other side of the island some penguins were bathing themselves in a tidal pool in the center of the colony. A hungry skua was walking around waiting for a chance to grab an egg from one of the nests. During the January and February hatching season these giant birds often take their chances and run a gauntlet of snapping beaks to prey on weak penguin chicks.

After about two hours on the island we boarded our zodiacs, motorized rubber rafts that cruise vessels use to ferry travellers from the ship to shore and for expedition cruises in the waters around landing sites. The adventure tour outfit we used is always keen on trying out new concepts and is known in the travel industry for finding new landing sites and taking trips other companies could only dream of. This time it was a crossing of the seven-mile wide Gerlache Strait in zodiacs, something that had never been done before. Speeding across the calm waters we cruised between icebergs of all shapes and sizes, fantastic sculptures left behind by the forces of wind and water. We witnessed a giant piece of ice break off a berg and tumble in a spray of ice-cold water into the deep waters of the Gerlache. One of our landings took place at Wilhelmina Bay, where we cruised around in our Zodiacs observing birds and wildlife and crossing fields of pack ice. We spotted a Leopard seal guarding a nearby penguin colony, waiting for its chance to catch an unsuspecting meal. Further down we observed a Crab eater seal relaxing on an iceberg as it floated in the current of the bay. We landed at the foot of an active glacier, a spot where presumably no one else had landed before. With just over 10,000 visitors a year chances are that at some point during your cruise you will be the first human visitor to set foot on a particular spot.

During our cruise we engaged in several invigorating hikes. Walks are optional and moderate and will take you to a penguin colony perched on top of a cliff, to a viewpoint overlooking the spectacular scenery, or exploring the remains of an old whaling station. No matter where you go the solitude of the land, the incredible vistas, and the feeling of being at the end of the world will profoundly strike you. On some of the cruises travellers can participate in sea kayaking or camp out on the ice. Kayaking is for experienced paddlers only as executing a perfect Eskimo roll is essential in the cold waters off the Antarctic Peninsula. Camping out overnight can be done by anyone and is of course an unforgettable experience. Complete silence will surround you while you feel like an early Polar explorer bundled up in the comfort of your sleeping bag.

We made numerous landings during our cruise in remote places like Trinity Island, Cierva Cove and Orne Harbour. Everywhere we went new discoveries and wildlife awaited us, from colonies of Gentoo penguins to flocks of Cape Petrel swimming near our zodiac; from a Minke whale crossing the bow of our ship to wandering albatrosses flying in the wake of the Akademik Ioffee.

Southern Elephant seals and Macaroni and Rockhopper penguins can be observed at several landing sites while acrobatic schools of Gentoo penguins can be seen making their way home through the deep waters off Deception Island.

On our third day in the archipelago we entered the caldera of Whale's Bay on Deception Island through the narrow straight known to sailors as Neptune's Bellows. Here we were to make our last landing before we returned across the Drake Passage to the wild lands of Patagonia. Just before arriving in Ushuaia we rounded Cape Horn in the perfect setting of a gale force nine-storm and 25 foot waves. Maybe it had been my whistling that brought on this storm, but it didn't spoil the moment for Allan Parker, an elderly Australian decorated with the Polar Star. He had been the leader of a daring rescue party, which saved the lives of some Australians who had been trapped in the icy Antarctic winter. It was at Cape Horn that we raised a glass for those sailors who did not make it through these turbulent waters and whose souls carried away to rest forever in the frozen continent, that last frontier. We felt as if a piece of our soul had stayed behind there too, in the solitude of the giant glaciers and snow fields; blending in with the cool evening air, swirling around forever over the sparkling blue waters of the Gerlache Strait.

I got off our ship early the next morning and stood for a while blinking in the early sunlight as the sun rose over the mountains surround Ushuaia. This must have been how Amundsen felt when he finally came home after his explorations of Antarctica. A lost soul wondering where to go next, fighting against the urge to turn around and make his way back towards the southern edge of Mother Earth.

Tom is with Trek Escape, Toronto.

The Fox and the Grapes

Carl Michener

It may sound unlikely, even unimaginable, but it happened… honest!

Being Canadian, I was surprised to stumble upon a beaver dam in the heart of the Patagonian Andes. (Hey, wait a minute… where am I again?) But there it was, right up near the tree line at 3,000 metres, a solid construction of partially debarked branches and thick peat. I sat down to look at the map, as it was late afternoon and beginning to drizzle.

It had been a tough day of hiking, but the visual rewards were as great as the climbs: thunderheads breaking apart and golden light streaming through, the fingers of God walking across green valleys and ruffs of snow cloaking the shoulders of mountains. Lakes shimmering far below like molten gold. You can' forget sights like those. At one peak's razor crest the wind was so strong that a few fragments of shale pelted at my cheek and drew blood.

As I sat in a copse of trees at the edge of this mountain meadow, I looked around and caught movement out of the corner of my eye. A beaver? No, a fox. A beautiful red fox, coming towards me. Coming closer. And closer. When it was five feet away I stood up and it skittered off, but not that far. Cheeky things, these Patagonian foxes! I walked over to my pack, dug out my camera, turned around and… hey! The fox was running away with my tin mug! I gave chase and it dropped the mug, slunk away, sat down and watched me.

As it turns out, there is a gray fox indigenous to Patagonia (foxes are called zorros in Spanish, isn't that great?) but both the red fox and beaver were imported from the Great White North (Canada, eh?) to kick start a fur trade. Patagonia's climate is not unlike that of northern British Columbia, and both animals took to the land like wildfire, damming mountain streams and stealing tin cups left, right and center.

I decided to camp for the night. Tent up, a little fire, a hot meal and a few cups of a particularly tasty Argentinean Malbec. I hung my pack from a tree, banked the embers, leaned the corked wine bottle against a log and hit the sack. I didn't hear anything throughout the night, but once I had rubbed the sleep out of my eyes the next morning, I saw that there had in fact been some activity.

It looks as if both a thief and a park warden had come to visit. The wine bottle was gone (much to my chagrin) and I couldn't find it anywhere. I imagined that sneaky fox chewing the cork and lapping up that purple wine. Even more amazing, the embers of my fire were scattered all over the place, and there were telltale scuffmarks where the campfire had been. Could it be that beavers, like rhinos, feel compelled to put out fires? I still haven't found out.

Carl Michener is with G.A.P Adventures.

Patagonia Hiker

The "Tibet' of South America

Red Hunt

Bolivia. What exactly does it have to offer? You could liken Bolivia to Tibet or Cambodia. They've each had their share of tough times and remain somewhat mysterious to most travellers. Tibet watches as China has its tourism soar, Cambodia watches as Thailand fills its streets with foreigners and Bolivia peers across the Andes to watch the tourists preparing for the Inca Trail in Peru. But that is the secret of Bolivia. Hidden in the shadow of its better known neighbour, this landlocked country can quietly carry on with its traditional Quechua and Aymara customs and keep to itself the secret it enjoys; Bolivia is perhaps the most awe-inspiring and beautifully scenic country in all of the Americas.

What makes Bolivia special? First of all, there are its people. Not only are they friendly, like in many other countries, but also their traditional Quechua and Aymara ways of life remain largely unchanged, even in the capital city of La Paz.

This country inspired the classic Conan Doyle novel *The Lost World*; it is home to the highest capital city in the world, as well as the world's largest salt flats (did you know beneath the salt is an actual lake!?). Vast deposits of silver once made the Bolivian city of Potosi the largest city in all of the Americas, and you can still take an eye-opening excursion to one of the nearby silver mines. Yet, somehow, Bolivia has been able to stay hidden from the large crowds of conventional tourism, letting the more adventurous of us enjoy the culture, cuisine, and stunning landscapes.

Did I mention how cheap Bolivia is? This is the place to be in South America if you want your travelling dollars to last a little longer. When it comes to great deals on the food front, fill yourself up with some "pique macho" a classic Bolivian meal of beef and sausage, with tomatoes, onions and various other ingredients, served with a heaping helping of fries. You'll find it served just about everywhere, and if you can actually eat it all, you usually won't have to eat again for days! Of course you could always try to go for some stir-fried llama if you feel so inclined?

So if you're heading to South America for a little adventure in Peru, or excitement in Brazil and Argentina, be sure to round out your trip with a stop in the "Tibet" of South America. Just remember, Bolivia is unlike any other country you'll ever experience, so here are a couple quick tips. Give yourself a couple extra days to explore and acclimatize if you fly into La Paz. There is some great hiking and some fantastic ruins nearby, but the altitude can zap your energy. Also, remember that at this height things are cooler. La Paz has an average temperature of under 10 degrees Celsius (50°F), so make sure you've got some warm clothes, especially if heading out on the salt flats.

TOP 5 LIST FOR BOLIVIA – Sean Benner, Operations Manager, G.A.P. Adventures

1. Incredibly rich and varied history and culture. Inca and pre-Inca culture and history is strong here, with Tiahuanaco being the oldest pre-Inca site on record. The indigenous people of Bolivia still are very prominent, and they count for a large number of the overall population (30% of Bolivians are Quechua). Tradition is maintained in much of the day-to-day life and is evident anywhere you travel in Bolivia.
2. Friendly people. No matter if you are in the bustling capital of La Paz, the scenic city of Sucre, or the tiny Andean mountain villages, the Bolivians are a welcoming people.
3. Diversity of incomparable landscapes and natural beauty. From the stunning to the bizarre, Bolivia's got it all. The Uyuni Salt Flats, multi-coloured mineral lagoons, volcanoes, hot springs, towering Andean peaks, the highest ski resort in the world, trekking, Amazon Jungle basin, the Pantanal, desolate deserts and more.
4. Wildlife. The Bolivian Amazon region is less explored than that of Peru, Ecuador and Brazil. This means that wildlife encounters and jungle experiences are top notch!
5. Adventure. Trekking in the Cordillera Real or scaling Huayna Potosi, the mountain experiences are unmatched. Other options could be rafting in the pampas, mountain biking, trekking and more.

Surreal Salt Flats: *Eileen Cameron, Central America Operations Manager*

The beauty and isolation of the landscape is what captivates me most. The first day we pile into 4 x 4 vehicles in the crisp air of Uyuni, in Southern Bolivia as we head out to the biggest salt plains in the world, over 12,000 square kilometres! In this world, 50 kilograms of iodized salt costs only two dollars. So much work for such little compensation, but that is life for many people in Bolivia. It is a strange feeling to be driving on this white salty crust when you know it is only a maximum of one foot thick and underneath is a saline lake. You may have heard of the "Ice Hotel", well out here in Bolivia, they have the salt hotel, which as the name suggests is completely made from salt. What an interesting site it is, the 'bricks' are cut from the crust of the salt plains and used for making walls and even beds. Luckily I'm not sleeping on a salt bed, as we head to San Juan, a small town of only 1,000 people. The enthusiasm and friendliness of the Bolivians is so evident here. Our group competed against the locals in a game of basketball, but after five minutes the G.A.P. Adventures team is keeling over from the exertion at 4,000 metres and the locals are running rings around us! However, it was a fun way to meet the locals through a sport that requires no verbal communication.

My Bolivian journey continues the next day, highlighted by seas of flamingoes and the Stone tree, made famous by Salvador Dali in his painting, 'The Persistence of Time'. But it's not long before the weather is what's on my mind. Up here it gets cold. Really cold, minus 20 degrees cold. But hey, when we come across a hot spring, our group jumps into the natural pool at 4,500 metres, looking out over a frozen lake with the Andes peaks in the background, talk about stunning scenery, it did help me forget about the cold, for at least a few minutes. My journey continued, with many more scenic pleasures, four days out on the Uyuni salt flats, what a memory. As an added bonus, during those four days we saw less than 20 people! Go to Bolivia, it is an oddity of a country by any comparison, but it is also the perfect place to get away from it all and ease your mind.

Red Hunt is an editor with G.A.P. Adventures.

Welcome to the Amazonian Philharmonic Jungle Orchestra

Dodie & Ronald Labbe

On the ninth of March, 13 ships and 1,200 men set sail under the command of Admiral Pedro Alvares Cabral, hoping to follow the footsteps of Vasco de Gama by navigating around Africa and reaching the passage to India. On April 22nd, they landed in Brazil, purely by accident. It was the year 1500, and the first recorded interaction between Europeans and Amazonians.

Five hundred years later, on a typical Winnipeg winter day, minus 30 degrees, my husband Ron and I left on a three-week journey to Peru, the National Geographic version. With an invitation from a friend who had been living in Peru for three years, we had jumped at the chance to venture to South America. To avoid the unfortunate navigation fate of Admiral Cabral, we booked and reconfirmed flights to Lima.

When Ron and I were first married, one of our top priorities was to travel the world. Circumstances dictated differently and our priorities were rearranged as we started our family first and travel plans took a far second seat. It was 17 years later and this was our very first ever winter holiday. We were pretty green and naïve but in retrospect allowed us to experience the trip of a lifetime.

We travelled to Lima and stayed with our friend for a couple of days acclimatizing to the tropical heat. From minus 30 degrees Celsius, no humidity, to 30 degress Celsius, high humidity, is quite a transition for one day. From Lima we flew to Iquitos (since the forest is so dense around the area you can only access it by air) and were met at the airport by the usual hoard of taxi drivers coupled with tour expeditions all trying to sell us their packages, in Spanish of course. We rode into Iquitos listening to the sales pitch and after talking to the tour company decided on an Adventure package. This included a ride up the Amazon in a speedboat to an "all-inclusive" resort. This amazing river got its name from Spanish conquistadors, who in 1541 found themselves in a fierce battle with women warriors fighting along with men. The strength of the women brought to mind the Greek myth of the Amazons, and the Europeans named the river Amazonas in their honour.

By the time we got to the resort (and I use that term loosely), it was dark. Here, it became pitch black shortly after 6:00 p.m., no dusk or twilight. As we pulled up to the dock we were advised to get our flashlights out. We hesitated, thinking that we really wouldn't all need a flashlight – until we were told that there were snakes in the grass along the path. We didn't need any coaxing after that. Stepping out of the boat, one of the guides shined his light toward a pole on the dock and showed us a tarantula about the size of the palm of a man's hand. Welcome to the Amazon!

Part of our package was an overnight stay at a camp away from the base resort. We hiked through the jungle for about half an hour to a tributary of the Amazon, where we boarded a canoe with our guide, Louie, and two of his helpers, along with our gear, food

Hut above the "Amazonian Philharmonic Orchestra" Dodie & Ron

and drinking water for the expedition.

Our friend in Peru was fluent in Spanish so we were depending on him for of all of our translation. But to our surprise, Louie conversed in English with ease. As we got to know him, we found out he had a university degree in botany and knew the names of all the plants and their medicinal purposes. His helpers were somehow related to him, but they spoke only Spanish. All we took with us was a knapsack with a change of clothes and what we needed for an overnight stay along with a mat for sleeping. This was a mat about three-feet wide, six-feet long and one-half-inch thick, which rolled up for easy carrying.

The Amazon rises high in the Andes, and rushes through gorges and a series of tributaries into basins in Peru, Brazil, Bolivia, Ecuador, Colombia and small portions of Venezuela. The river is very wide and straight and makes its way through various terrains.

For the first hour or so while we paddled, we passed through thick trees and overhanging branches, breathtaking to behold. Once we reached the open water our guides started the small motor and we motored along the tributary in the "black water." This inky water, they assured us, was not polluted, but was black because of the vegetation that had fallen in.

There were actually three basic water types along the route: white water, clear water and black water. The black water was like black tea, due to the discharge of the Urubu River, a tributary. Unlike in the clear water, the rocks in the black water were so old and had been washed over so many times that they had been leached of minerals.

We passed many other boats along the way that were bringing produce into Iquitos to market. We passed a few villages and then stopped to visit one.

The people we saw along the tributaries of the Amazon, called river people, lived by farming coconuts, bananas and other tropical crops that grew in the rainforest. After we motored slowly for a couple of hours, the tributary started getting narrower and shallower to the point when we finally docked we couldn't have gone any farther on the river.

We arrived at a typical home in the area, a hut by Canadian standards, made of bamboo and built on stilts, where we were to stay overnight. On the upper level was a veranda, a small open kitchen on one end and one walled room used as a bedroom and storage area. On the ground, under the house lived animals, such as turkeys, chickens and dogs. (No anacondas slithering around!) Well, as long as we weren't all sleeping together, I felt comfortable.

We were welcomed into the home by the owners, who seemed to be more relatives of Louie's. They happily gave us a tour and were quite proud to show us the paka, a river rodent they had caught and chopped up to make into soup. They offered us a piece of smoked alligator they had cooked the day before. Coming from the Prairies and raised on the farm, the only culinary diversity I was aware of was potluck. As much as I wanted to try this cuisine and not offend my hosts, I just had to pass on the alligator. Ron, however, was brave enough to try the alligator, which he described as rubbery and bland.

The man of the house took us for a hike through the jungle that was considered his property. The trail was deep, but clear - not the type you need a machete to cut through, as we had anticipated. We walked safely at all times with gumboots on and the buttons on our long-sleeved shirts all buttoned up. We saw a gigantic cedar tree that was being cut up and sold for lumber. We also were given a demonstration of how palm fruit is harvested.

After our hike through the jungle in above 30 degree temperatures with high humidity, we went back to the camp. We were so hot that both Ron and the friend we were travelling with took a dip in the Amazon after being assured that there were no piranhas in the water. I was so hot and sticky that I couldn't find the energy to strip my clothes off and try to get my bathing suit on. I was also a little nervous about the piranha issue.

After the swim we had supper – fried Spam, an egg and bottled water. It was delicious.

By this time it was dark and time to set up our sleeping quarters on the veranda of the hut. Each of us had a little tent made with mosquito netting just the width and length of our sleeping mat. The floor of the veranda was made of bamboo poles lashed together. Everybody went to bed shortly after supper because it was too dark to do much else.

This is when the fun started. It was still very hot and there was no breeze that night. Being inside the little tents was a bit claustrophobic so we were a little uncomfortable in that way. As I mentioned, the sleeping mats were very thin and you could feel every little ridge on the bamboo floor underneath. We had no pillows, so we used our knapsacks and spare clothes as pillows.

We had just started dozing off when the tree frogs started their chorus. The one piece of technology that we wished we had brought with us was a tape recorder because I cannot describe the noise that these animals make. I have never seen a tree frog but I imagined them to be the size of a large cat from the noise that they were making. We lay awake listening to the croaking for at least two hours.

Once the frogs finished, we listened to the other noises of the jungle, including all the animals under us that were supposed to be sleeping. The poor mangy-looking dog kept trying to sneak up on to the veranda and we were forced to throw our shoes at it to keep it off. We dozed on and off through the night, waking up to check our Indiglow watches to discover that we had slept for another five or ten minutes every time.

At about 3:00 a.m., after the dog settled down, the turkeys started up. Every river hut had a turkey and all the turkeys knew one another. You've heard of the Africans talking with drums? Well, in the Amazon the turkeys talk turkey -- continuously, from 3:00 a.m. to sunrise.

We didn't get much sleep that night, as we were the all - night audience of the Amazonian Philharmonic Jungle orchestra!

Did I happen to mention that there were no bathrooms or outhouses? I don't know what the residents did, but we found a tree out of sight of the hut and did what we needed to do. When we left the camp to go back to the main resort, all I wanted was a civilized shower and it was the thought of this piece of heaven at the end of the day that gave me the energy to keep going.

We loaded up the boat and started back only to discover that the engine was running low on gas. Sure enough, about an hour into the trip back the engine stopped and that is when we found out that there were no paddles in the boat. Who coined the phrase, up the creek without a paddle anyway?

At this point it started to rain. It was a typical warm and heavy tropical rain, a five-minute downpour that created havoc and then stopped like a dead-end road. After pulling out our rain gear, we grabbed pots and pans and anything in the boat that would serve as a paddle to get us to a village where we could buy some gas. We did manage to get enough to get us back to the main resort.

We pulled in midafternoon only to find that the water system was not working, so my dream of a shower was postponed until we got back to Iquitos. That night we splurged on a well-deserved, four-star hotel with hot and cold running water and a big comfortable bed that needed no invitation to flop on.

The trip down the Amazon was enough of an adventure, but it was just the first leg of our four-week trip. From there we headed back to Lima, then to Cusco. We took a three-hour train ride and hiked Machu Picchu and then the 12 hour train rides to Lake Titicaca.

Before this trip was even over, Ron and I were already saying to each other, "Next time lets take a conversational Spanish class when we come back" and "Next time, now that we're in our 50s and we're certainly not Amazonians, lets splurge for a hotel within our comfort level!" We have had many other travel escapades since this wonderful trip but this by far was the most exciting experience we had undertaken in our travels and an adventure story to be retold.

Dodie and Ronald Labbe reside in Winnipeg, Manitoba.

Tied Up with String in Equador

J. Case

In the days before 9/11, security was much looser than today. During my career as an environmental consultant, I have travelled throughout the western hemisphere. One week, I would be in snow covered Aklavik in northern Canada and the next I would be in Bogotá, Colombia. Often this travel required going on some of the more "colourful" smaller airlines.

On one occasion, while waiting to fly to Ecuador, I was watching passengers check in for the flight to Quito. Many of the folks where returning home to South America after a visit to the United States and were obviously taking with them gifts or purchased items they could not get at home. While most people carried souvenirs like ghetto blasters with dual tape drives, one gentleman arrived at the check in counter with a brand new chainsaw. A lively conversation ensued in animated Spanish between the would-be passenger and the counter attendant. She informed him in very polite "usted" Spanish that on no account would he be allowed to carry an unwrapped chain saw onto an international air flight. Appearing somewhat puzzled the fellow wandered away, ticket in one hand and chain saw in the other.

Minutes later we started lining up by seat number in preparation for boarding the plane. As I was walking onto the plane I noticed our lumberjack arrive back to the check-in counter with a big smile on his face. Clutched in his arms was obviously the chain saw, now carefully wrapped up with newspaper and string. The attendant took his ticket, gave him a boarding pass and we all flew happily into the night.

When you travel,
remember that a foreign country
is not designed
to make you comfortable.
It is designed to make
its own people comfortable.

Clifton Fadiman

Rick's Café, world famous Cliffside haunt, Negril, Jamaica. We spent our honeymoon watching the brave leap off the cliff from a 33 foot high platform. The cliffs boast a spectacular Caribbean sunset that is truly magnificent & unrivaled.

Jeannie & Larry Henke

L' Atitudes

Louise Hudson

Part One

Departing from our Calgary suburb we are now into the first month of a round the world educational trip on a Semester At Sea ship. Surrounded by 700 university students, my husband Simon is teaching tourism and business and our two sons, Rupert (13) and Fergus (10) are attending a floating, international school (somewhat different to their usual Canadian equivalents).

Our voyage, with stopovers in ten different countries, is four months in total. Students from the US, Canada, Israel, and the Far East have opted to gain three academic credits towards their undergraduate degrees in subjects ranging from law, to theatre, to literature whilst travelling the world.

Starting from Nassau, Bahamas, we are visiting Puerto Rico, Brazil and then crossing the Atlantic to Cape Town, South Africa. We then visit the tropical paradise of Mauritius before docking in Chennai, India. The politically controversial Myanmar (formerly Burma) is our next port of call, followed by Vietnam. Hong Kong, China and Japan are our Far East destinations and then we make the two-week sea crossing to San Diego, just pausing to refuel in Hawaii.

We flew from a minus five January day in Calgary to balmy 80 degree (in US Fahrenheit!) Fort Lauderdale and then on to warm and wet Nassau. This is my account of our highlights and mishaps so far on the first leg of our four-month voyage.

With two Caribbean countries and nine days at sea under my belt, I'm feeling like an old seadog. I haven't needed any pills or patches so far, but the first three ocean bound days with a lashing wind and a swell of 15 feet required much sleeping and even more fresh air, sunbathing and exercise on the top deck to counteract the incipient nausea. Both Rupert and Simon succumbed to sickness but gradually acclimatized with the help of medication and rest. I don't think any of us expected such heavy seas or anticipated the challenge to core stability of mere walking, showering and sleeping, let alone any kind of work or study!

Nassau, Bahamas was our suitably exotic departure port, and we languished there for five days on a postcard caliber white sandy beach resort before boarding our mobile home for the next four months. You can picture the scenario, courtesy of all those locally-filmed James Bond movies: perfect sandy crescent bays fringed with wind-animated palms, shark-infested turquoise seas and violent Crayola sunsets, preferably viewed from a flower-bedecked patio with a pina colada in hand.

Something you may not have heard about here, though, is the vicious competition, in the famous Cable Beach area, between hotel magnates desperate to outdo each other. Atlantis Resort, a Las Vegas-style underwater-themed monolith, is an already firmly established eyesore on Paradise Island. Its outsized, tacky opulence has prompted the new, oil-rich owners of Cable Beach to go one step further. So watch out for the US$2 billion Crystal Palace, which will be built over the next five years. Even the contours of the famous, perfect beach will be altered to fit the designers' plans. Naturally, as in Vegas, casinos are helping to justify the vast investment but at what cost to the environment and wildlife, let alone the ingenuous tourists' savings accounts?

Hairstyles, believe it or not, were a notable feature of Nassau! Bahamian women (and some men, too) spend much time, money and ingenuity on the most amazing updos. Only the tourists these days indulge in the Bo Derek-style braiding and beads. The locals have moved onward (and upward) with incredible, smoothly coiffed, architectural works of art on their heads. How do they sleep? According to Lisa, my hair braider (sorry, had to do it just this once in my life), these glamorous women spend US$70 or more every week or two for a new hair creation and manage to sleep and shower and still keep them intact.

Flamingoes, the brilliantly flamboyant national bird of the Bahamas, were another memorable native feature. We only saw them in captivity (although there are 50,000 in the wild on the mainly national park island of Great Abaco) but were vastly amused to watch their military antics as their trainer put them through their synchronized paces on the parade ground at the Ardastra Gardens Zoo.

An introduction to scuba diving was a must for my two sons, so we booked a US$108 package with Bahamas Divers for a morning's pool instruction followed by a real dive on the reef. The pool time seemed awfully short to me and the teaching somewhat scanty, but, nevertheless, we boarded our boat after lunch along with about 30 other beginner divers and snorkelers and headed off 30 minutes out to sea. We moored at a sunken wreck and blithely the boys, and everyone else, departed with their guides, leaving me to worry on board (about sharks, oxygen failure, electric rays, man-eating squid, etc).

To pass the time, I dived into the irresistibly blue waters and, avoiding rock, reef and coral, swam nonchalantly with head well above water, pretending there was nothing in there. When our boat boy retrieved a spare snorkel from a hyperventilating fellow passenger, I was reluctantly cajoled into trying snorkeling despite my admonitions that I really didn't want to see any fish. Typically, I was the only one that day to find anything even remotely scary. Having, amazingly, got the hang of breathing through my mouth underwater against all my natural survival instincts, I carefully veered away from other snorkelers in order to avoid collisions which I knew would have inevitably resulted in horrid internecine struggles. Big mistake. After happily viewing pretty but outsize aquarium fish from a safe height ten feet or so above, I suddenly came face to teeth with a nasty, be-finned, silver predator. It was scything through the water towards me just below the surface at perfect snorkeling level and was obviously not inclined to give way. I am still astonished (and gratified) that when confronted with the essence of my worst nightmares, I am inexplicably endowed with Olympian swimming skills during getaway! Apparently it was "just" a barracuda but they are all sharks to me.

Jonkanoo culture is a colourful feature of the Bahamas and we sampled the native food

– notably rubbery conch fried and battered in various sauces – as well as the music. We were thoroughly entertained on the Friday night in the Johnny Canoe restaurant at Cable Beach by a surprising procession of grass-skirted and ornately head-dressed native men belting out their jangly music on a combination of trombones, trumpets, various drums and handheld instruments. They appeared at regular intervals all through dinner, marching between tables in their sequins and feathers, providing much needed light relief from the 50s and 60s crooning of the live band.

Being winter in the Bahamas, some of our beach days were interrupted by cloud and rainstorms and we had to resort to the hot tub next to the pool to warm up. Temperatures ranged from around 20 to 25 Celsius, though, so a welcome respite from Canadian coolness.

The prevailing atmosphere in Puerto Rico was so entirely different to the African-influenced Bahamas. In Nassau, it was "taxi, taxi" or "hair braiding" cries from every street vendor, with colourfully clad kids offering "a nice song for a dollar" and reggae style music belting out from the bars. In Puerto Rico's capital San Juan, tiny cobbled streets with crazy traffic, ancient buildings, wrought iron balconies, quiet, hidden courtyards and bougainvillea-swamped archways, were redolent of Spain for me. Here it was salsa and bomba music emanating from each doorway and sangria replaced the pina colada. The two well-preserved forts and the imposing stone walls encircling the old gated part of the capital epitomized. Four hundred years of colonial history. Puerto Rico was the gateway to the Americas favoured by European troop ships and profiteers. The Spanish conquerors on the island had a tough job fighting off the English, Dutch and French invaders so they resorted to building 250 years of fortifications and walls to keep them at bay.

Our trip highlight was the native dance and music festival, which is held every Tuesday night at 6:00 p.m. at the San Cristobal Fort. With the crumbling grandeur of the ancient fort as a backdrop, twilight slowly descending, and the cooling evening temperatures, this al fresco display of vibrant and lively professional dancers, musicians and singers captivated us. Even the kids interrupted their games to watch in fascination. A quarto player enraptured us with the beautiful, tinkling, banjo-like sounds emitting from his small, guitar shaped ten-stringed instrument. And the bomba drummer exhausted himself with incredibly deft and flamboyant manual dexterity, creating an incredible array of sounds from just one drum and synchronizing all his actions perfectly with the dancers.

Finding a five-inch long stick insect was a real achievement during our rain forest excursion. El Yunque Rainforest is a national park easily accessible from San Juan, full of very distant birds hiding in the emerald canopy and famous for its tree-dwelling two inch diameter snails. The rainforest is huge and really necessitates a guided tour. One hiker got lost there recently and finally made it out after two weeks terrified wandering! Coca Falls provided a wonderful opportunity to strip down to bathing suits and immerse in the cool, fierce gush of the waterfall. Great photo opportunities too in between the rain showers!

Luquillo Beach is beautiful in the picture postcard tradition: long, white sandy shore, azure sea teaming with fish, and palm trees hiding a quaint strip of small stores selling fruit and fast food. But the town beach was the biggest surprise. Just four blocks from the Navy Frontier Pier where we were docked, we walked across a lovely, shady park to the locals' beach with its spacious car park and (thankfully) high-rise-free shoreline. The water was encircled by a reef protecting it from the surf and perfect for swimming and snorkeling if you don't mind sharing it with a four foot long barracuda! Yes, we saw another one – this time it was Simon and Rupert so now they know how frightening they are. The foliage is so lush here that even the beach is dotted with sprawling ivies and pretty pink flowers.

Imagine how hot it must be here in the summer! We were frazzled just walking around the old town in January at 28 degrees and had to resort to regular piraguas to cool down. Piraguas are the Puerto Rican version of snow cones and are intermittently available at street corners for US$1.25 in many flavours. Other culinary highlights included churrasco, a skirt steak that Simon had to prevent me forcibly from ordering three nights in a row! It is a succulent, marinated, long, narrow piece of steak, often served rolled up for decoration, eaten locally with rice and beans. The town was suffering from its post-fiesta weekend hangover when we arrived on the Monday so we had some initial trouble finding lively restaurants, but the Café Puerto Rico, in the Christopher Columbus Plaza, was always busy and the food and service were equally high standard and good value. In fact, prices in San Juan were noticeably cheaper than in the Bahamas – more commensurate with Canada.

Well, guess what? After all that sightseeing, culture, hummingbird spotting and sun soaking, we really didn't want to get back on our ship! This was partly to do with the dread of seasickness as our first voyage from the Bahamas was incredibly rough and had made even the seasoned ship staff ill. However, ship life is exciting for us all with its huge variety of activities, educational facilities and luxuries. Our cabins are spacious and well appointed. Simon and I even have a balcony complete with chairs and table from which to dolphin-spot or contemplate the sea-bound terns performing their thermal dances and audacious dives after abundant fish churned up by our ship. The food is very good particularly when accompanied by lively conversation outside on the deck stargazing. We have intellectual, musical and social soirees to while away the evenings and, of course, the daytimes zoom past in a combination of gym exercise, lessons, meetings, presentations, games, swimming and sunbathing.

Our first real interaction with faculty, staff, crew and students was provided by "Neptune Day": the "ancient" celebration of crossing the Equator. Classes were suspended on January 30th for this cultish event whereby a robed and bearded Neptune (the Executive Dean) proceeded through the ship's corridors with his courtiers and a band playing tribal music on drums and whistles awakening all at 7:30 am. Initiates were then dunked in a viscous fish-gut concoction, made to kiss a fish and "baptized" in the swimming pool or volunteered, instead, to have their heads shaved. Both our sons, contrary to everyone else on board, actually enjoyed being dunked and returned three or four times. And guess who also had his head shaved? Our younger son, Fergus, who surrendered his thick, shoulder-length locks to be one of the "skinhead club" which comprised at least 30 guys and girls by the end of the day!

Part Two

Bums, butts, bottoms and buttocks will definitely linger longest in my Brazilian memories, differentiating these uninhibited people from any other culture. The body is revered here and whether young or old, fit or fat, male, female (or any mixture thereof), it appears to be your social duty to clothe it scantily and provocatively and strut it on the beach or street. Can you imagine guys striding purposefully through crowded urban Salvador scenery attired only in diminutive Speedos sensibly accompanied by practical socks and shoes? It wouldn't work in Superstore, but in Salvador no one (except slack-mouthed tourists, such as myself) pays it any heed. Equally, ladies of all shapes, sizes and ages sport the famous Brazilian-cut bikinis, baring their bronzed buttocks brazenly on the beach; in outdoor cafes and even in a seashore turtle sanctuary we visited.

Salvador is a wonderfully vibrant mixture of ancient culture, music and religion plus up-to-date technology, industry and materialism. In the old, upper town of Pelourinho, buildings co-exist in an alternating pattern of run-down and renovated. Reached from the dockside by a high-speed elevator, decadently embellished baroque churches and cathedrals overlook vast plazas, commemorating city founders, where hordes of pigeons vie for pavement cafe leftovers. Sinuous cobbled alleys lead scarily down into "favelas" where people live in windowless squalor. Just nearby there are modern malls, Internet cafes and gated grand villas, juxtaposed with sleazy slums.

The "Candomble" religion is kept very visible today on the crowded city streets by afro-origin women in gigantic, white, lacey and frilly frocks who sell "orixa" beads (different colours for the various gods and goddesses) and cook local gastronomic delicacies based on bean flours and shrimp. Some, whose abject poverty counteracts their religious ethics, offer themselves instead if they can't quite close the sale! A favourite sport for such street-smart sharks, often cunningly disguised in this virginal attire, is to beguile unwary tourists into taking photos of them and then demand payment. We refused to pay up after being cajoled into posing with three, forcibly insistent women and the row, in Portuguese, would have been quite entertaining to recount if we could have understood any of it!

Capoeira is an ancient martial art, brought over originally by African slaves, and kept alive by artful concealment from owners and religious leaders during hundreds of years of servitude. In Sunday church services, Candomble was secretly superimposed by slave practitioners onto Catholicism and, similarly, African male slaves turned their Capoeira aggression into a symbolic dance accompanied by drums and a bow-and-bowl instrument whose sounds resonate against the player's stomach. Again, one problem with watching and photographing seemingly spontaneous street performances is the resultant exigent demands for money.

Pelourinho (meaning the "whipping post" where enslaved Africans were initiated into their new life of servitude by the ruling Portuguese landlords) is now a tribute to the infusion of African culture into everyday Brazilian life with its sensuously insistent drumbeats, intricately beaded hairstyles and jewellry, brightly coloured clothing and rhythmic tribal dancing. In addition, the Mediterranean style alfresco dining and people watching from street cafes make both daytime sightseeing and nightlife exciting experiences. But, you have to add to this a pervasive sense of insecurity and anxiety as distracting street performances and tightly packed crowds offer ideal opportunities for street crime.

Desperation has lead many poverty-line city dwellers to stealing, pick pocketing, mugging and violence. One student was held at knifepoint in Salvador until he surrendered his cash. During our visit, another girl had a gold chain wrenched from her neck, and a professor was relieved of US$10 by three knife-brandishing thugs whilst on a trip to Rio de Janeiro. After many warnings in our paranoia-inducing "Pre-port Briefing". I spent much of my city wanderings clutching my youngest son in sweaty hands in order to protect him from molestation or kidnapping into the white slave trade or death by red-light hopping, speed limit exceeding traffic. Drivers often skip the lights in Salvador in order to avoid carjacking and, apparently, at night this is perfectly legal. After a fabulous dinner out at Mama Bahia's in the upper town, our taxi driver demonstrated this local legal loophole, running three red lights consecutively in order to get us "safely" back to our ship. We had been told not to walk for fear of knives and guns, but it seemed to us we were in as much danger from seatbelt-free taxis.

For me, the Salvador experience was somewhat marred by its overtones of poverty and crime. Street and beach vendors also had no respect for personal space or comfort zones. An innocent conversation with a jewellry salesman on the city's Farol do Barra beach resulted in a flock of motley vendors descending like vultures upon a girlfriend and me. After several requests in pseudo English/Spanish/Portuguese for them to leave, accompanied finally in desperation by the kind of "shoo-ing" actions that usually work with their avian counterparts, we were treated to a volley of abuse. Strange how they knew all the English swear words!

Churrasco has become a meat lover's all-you-can-eat banquet here and Boi Preto Restaurant near the airport is the epitome of its hedonistic delights. There you can indulge in every meat (and I mean every meat) all in one meal accompanied by an orgy of lobster, crab, oysters, outsized prawns, mussels, Swiss cheeses, foie gras pates, various potatoes, soups, salads, rice dishes, an array of sushi and fruits all for a set fee of US$25 per adult (US$15 for kids). The lamb, pork, beef and poultry are served constantly from machete-brandishing waiters who slash off as many portions as your over-extended stomach can deal with. It reminded me of the gigantic python who injudiciously swallowed a huge alligator in Florida and then proceeded to explode. You are even provided with a diagram of beef parts so you can identify which is rump and which is hump, making this bacchanalian feast the ultimate vegetarian's nightmare. Ironically, the bus ride to the restaurant took us through some of the poorest neighbourhoods of the city and made us hope that at least the leftovers are donated to the starving.

The hour and half journey north to the middle-class beach resort of Praia do Forte was a huge sigh of relief to all the family. Whilst enamoured by the bustling diversity of the colourful city, the boys had been somewhat traumatized by the menacing urban aspects of Brazil, imbibing fear and insecurity from us, no doubt. As we drove through the sprawling suburbs even the hillside slums looked charismatic in their terracotta simplicity, swamped with lush foliage. The earth is brilliantly red around Salvador and lends itself to a charming rustic pottery and ovenware industry. For just three "reals" (about US$1.50) you can pick up a sturdy, ovenproof serving dish.

Praia do Forte is famous for its turtle conservation project. It is a great example of successful eco-tourism whereby the local fishermen realized they could make more money out of preserving the endangered turtles than by hunting them for their meat, eggs and extravagant shells. Around the reserve has developed the most classy and quaint village, sporting an immaculate church and one-room school, protected by security guards and gates. The river and rainforest are also conservation areas penetrated only by organized trips in jeeps or kayaks to view the rare birds, foliage and monkeys. You do have to be lucky to see the monkeys but one fellow Semester At Sea passenger was treated to a monkey tea party at a restaurant when a bunch of them climbed down for a fruit feast provided by the waiters.

It's the sort of place that many tourists come for a day trip to see the turtles and the rainforest and wish they could stay longer. Luckily, some friends had tipped us off in advance that it was a small slice of Paradise, so we had booked a "pousada" (guesthouse) for the weekend. Pousada Dos Artistas was utterly enchanting with its wood-trimmed adobe construction, hammock-kitted balconies, rattan furnishings and beautiful artwork, all set in a privacy-ensuring lush landscape of giant trees, ferns, creepers and unreal-looking flowers.

Simon and the boys snorkeled in an array of low-tide rock pools, avoiding deathly stonefish and revelling in the bath-temperature underwater beauty. I declined after having the serendipitous opportunity to study the teeth of a (fortunately dead) barracuda speared by local divers just a few metres offshore right in front of me. Sardines were also being dredged up in wispy nets in the space of a few minutes by enterprising youngsters in row boats or windsurf boards and were then rushed to the beachfront cafes to be cooked whole in a light batter to gratify the lunchtime appetites of Brazilian and Italian tourists. I say Italian, specifically, as it is a favourite wintering destination for Italians craving yet more good weather. In fact, every other restaurant in the exquisite, cobbled, car-free Main Street specializes in Italian cuisine and ice cream.

My favourite local delicacy was "bolinhas do queijo" which are mouthfuls of hot, melted cheese fried in a battered ball. "Bolinhas do peixe" (fish balls, not pixie balls) were another hit with the boys as a lunchtime snack. We also had wonderful, home-cooked tortillas as a nacho-style appetizer and all sampled the local sardines and line fish.

The difference in Praia do Forte, apart from the lush countryside, low rise, pristine buildings, sparklingly-kept shops and restaurants, immaculate toilets, lack of poverty, etc, could be summed up by the safe knowledge that if you dropped your wallet here, someone would for sure come running to return it to you rather than pounce and sprint in the opposite direction. And they wouldn't even expect a reward apart from thanks. Our kids were safe to roam unaccompanied. Vendors were polite and respectful of personal space. If only the hierarchy in Salvador could recognize that where industry is lacking, tourism is a fabulous way to create employment, clean up a city, preserve its culture, and give its poorer population purpose, a livelihood and self-respect.

Football is an absolute must for any visit to Brazil. It is "the opiate of the masses" and Brazil is, of course, traditionally one of the foremost forces in international soccer. And what a different atmosphere to its European counterpart! Even the group chants sound rhythmic and tribal with the ubiquitous drumbeats adding a dancing, light feel to the male-oriented audience. Despite horrific toilets, intense heat even in the shady part of the stadium and uncomfortable concrete seats, the experience was totally positive although the standard of soccer in the Bahia region at the moment is not quite World Cup with Pelè at the helm. Having a British soccer-obsessed husband as well as a fledgling ten year-old addict, even I know a thing or two about the game. It was rather too frequently interspersed with dirty fouls, yellow cards and enthusiastic stretcher-bearers but was enlivened by very vivacious crowd support, frequent fire-crackers sounding ominously like gunshots, bats scudding the pitch, and con-merchant vendors selling every kind of drink, ice cream and snack.

Unfortunately, exposed to so much alcohol, cheap and freely available (until they actually drank the bar dry), many of the students overindulged and risked life and limb clambering up and down the cement stairways in frivolous flip flops. This footwear was also a potential health hazard in the overflowing toilet areas. Over 400 faculty and students attended this match, mainly cheering expediently (since we were in the home enclosure) for the local team, Bahia. Shepherding this motley crew onto the eight buses at match end was compared to a war zone military extraction by our fearless photographer who managed to rescue stragglers caught up in the maelstrom of drunks, armed police, and crazed fans. It never felt dangerous, though, until after the 1-1 match when football fans suddenly appeared to be spoiling for more excitement and the "filets mignon", as street sharks call us ignorant tourists, were irresistibly abundant.

However, we all managed to get back to the ship and everyone was safely on board by the 9:00 p.m. curfew. If you are late you are docked time, which is then multiplied and given back to you as a time penalty. This means that you have to stay on board extra time when everyone else has disembarked at the next port! This strictly maintained rule has led to the new blood sport of heckling from the top deck as stragglers limp, run or hitch rides in a desperate attempt to beat the clock.

Ship Life

What's so blissful on board is that the phone doesn't ring. You can get on with projects without constant interruptions and mundane concerns are suspended for the whole voyage. Another strange thing about endless sea travel, with no sight of land, other ships or civilization, is that there is, contrarily, no time for boredom.

Superimpose a university campus (studious by day, raucous by night) onto a luxurious, full service cruise ship and you will be able to envisage the daily life of Semester At Sea. In order to gain access at each embarkation, every voyager has to present an electronic ID card, have their bags checked for contraband (after Brazil several pygmy blow-pipes were confiscated) and students are frisked for illicit alcohol ingeniously strapped to their limbs. Our Kremlin-trained customs team assiduously sniffs even students' water bottles.

Students' alcohol intake onboard is strictly monitored by a ticket system; resulting in a boozy frenzy each time they touch land. "Man over board" is definitely deterred by strict policing of the student parties and limitations per person per night on alcoholic consumption. Luckily, this policy is not extended to the out-of-student-limits Faculty Lounge where between 5:00 and 6:00 p.m. and from 9:00 till 11:00 p.m., staff, senior crew, professors and spouses can indulge in some alcohol therapy after the daily grind. However, the ever-present threat of seasickness is a kind of limit in itself.

Our revered and feared Captain and his fellow officers sometimes join the Faculty bar flies in their pristinely-pressed white uniforms and that's when we all realize that the forebodingly austere Croatian is actually one of us humans after all. I still find it amusing, though, to see him in shorts or jeans on his days off.

We have become a family on board already with total freedom to chat with anyone from staff, to students, to crew, to administrative personnel. Most voyagers seem to be quite extroverted and fun loving, keen on lively conversation, social and intellectual group activities and sharing their interests and hobbies with each other. Sport and exercise are difficult to accommodate on our crowded decks but our gym is quite well appointed (booked solid from 6:00 a.m. till 11 p.m. on the sign-up sheets!) and there are organized yoga and kickboxing classes as well as volleyball, basketball and soccer games. The keeling of the ship offers a new dimension to scoring, of course. And, unfortunately, during our rough sea spell, over 50 ping pong balls were lost to the waves on the first day of table tennis playing so that pastime was suspended for a while.

Of course, with seven decks, we have lots of stairs to climb up and down, but still most people are bemoaning their increased girth after a month of vacationing and self-service, buffet-style over-indulgence. The food is abundant with at least three choices for each meal plus a variety of salads, fruit, cheeses and desserts. It's a mixture of international and American favourites with peanut butter and jelly available for US comfort-eaters. Some people complain that it is a bit bland and repetitive, however, and lament that there's no wine to sip whilst lingering over dinner, contemplating the sunset or the stars.

Days are separated into A and B days, completely ignoring weekends and further disorienting us in our tenuous associations with the real world. We also have to jump an hour ahead regularly to keep up with the time zone changes as we head east. It can ruin the party to contemplate an hour's less sleep with 8:00 a.m. classes the next morning! Simon teaches at 8:00 a.m. on A and B days (having requested mornings but never contemplating such early starts!) with a total of four and a half hours of teaching over the two days. This is the same for all students and professors who also have to fit in their prep work and marking – preferably up on the sundeck which is jam-packed from 10:30 a.m. each day.

Global Studies is a mandatory class for students and faculty daily between 9:30 a.m. and 10:30 a.m., so that's the time I work out and craftily bag my lounger by the pool. We all receive a memo each day from the Executive Dean with details of activities, meetings, administrative and immigration procedures, problems and warnings, upcoming social events, etc. And the Assistant Executive Dean amuses the entire ship twice a day with his semi-jocular but utterly serious admonitions over a very loud speaker. "We are two tampons short of a toilet crisis." bellowed at noon throughout the crowded dining rooms totally brought the house down.

Profs, spouses and staff have the luxurious retreat of the Faculty Lounge at the front of the top deck with panoramic views of endless sea and resplendent sunsets. Presumably you can also see dynamic sunrises but, for me, that's out of the question at 5:00 a.m.

Now we are one huge extended happy family, we celebrate everyone's birthday with gusto. Huge ice cream cakes, candles, balloons, parties and door decorating make certain that these almost daily occasions do not pass unnoticed. Tonight's 50th extravaganza for our resident video-maker is providing the perfect platform to make an utter fool of oneself with its karaoke, music and dance theme. Let's hope the tequila is flowing sufficiently to eradicate academic inhibitions!

Onboard we also have a senior group who participate in their own community college as well as the general lectures, global studies, organized land field trips and the social scene. They also act as surrogate grandparents for homesick students. Many of the female members are joining in a quilting project which involves collecting fabrics from each country we visit, throwing them into a fabric pool and then benefiting from all the different choices, making squares or hexagons and sewing them together into a patchwork quilt, wall hanging or pillowcase to commemorate our voyage. Guess what, I'm joining in too! Can't wait for Africa and the Far East where the fabric choices will hopefully be myriad and cheap.

Big Brothers and Sisters and the Extended Family Program have also been set up for the children onboard. Fergus's new brothers are keeping him busy in his free time with soccer and basketball and Rupert's teaching him poker and the guitar. The eight children attend school for four hours per day with an optional fun hour in the evenings for games. They all bonded quickly into a pack after one day of shyness and now can rarely be located for such boring things as teeth cleaning and bedtime.

Mainly from the Philippines, our crew is a total joy. Both males and female, cabin staff and restaurant/bar personnel are immaculately turned out, trained in topnotch service, and all vivacious and friendly. Antoinette, our extroverted Faculty Lounge bartender has an infectious zest for life, dancing and singing during work, despite having had to leave her seven-year-old daughter back home with family.

Whenever we are nearing our next port of call, we have mandatory pre-port lectures. The first, dubbed the Pre-Port Panorama by me, is an overview of the history, culture, language, sights and sounds of the specific city and country approaching, featuring indigenous lecturers who ride the ship for just that one leg of the voyage. The second, Pre-Port Paranoia, contains lurid warnings about crime, medical dangers, and typical tourist faux pas, which could easily deter the timid from disembarking at all. Our ship's doctor regales us with vivid descriptions of the bone-breaking agonies of dengue fever death, the mucky minutiae of traveller's diarrhea, the certainty of rabies from animal-petting, and the high risks for traffic deaths whether inside or outside the local cars. And, for a finale, our ex-police officer security guard recreates previous Semester At Sea crime scenarios, urging us to carry "throw-away" fake wallets, wear our credit cards in our shoes (except in Africa where they may want your shoes more than your Visa), and never to resist gun-totting muggers.

Apparently, in the more remote and poorer areas of South Africa, it is wise to carry a spare bra or pair of underpants for bartering for art or craftwork. Scanty panties from La Senza or Victoria's Secrets are far stronger against the rand than the dollar, it seems. Not sure if my teabag-dyed (skin-coloured to avoid VPL) g-strings will qualify but at least they fit nicely in my wallet.

Whilst on board Simon and I are writing up dozens of "case studies" garnered from interviews and information from each city and centre of tourism we visit. Naturally, we get the majority of the work done during cloudy days. Temperatures during this trip across the Atlantic from Brazil to South Africa range from 30 degrees Celsius in the day to a steamy 25 at night but it's often very windy on deck.

Stargazing is a gratifying nighttime sport when cloud-free and we have our own astronomer onboard to aid identification in this upside down southern hemisphere. We also have our own ornithologist (who is now famous for spotting "masked boobies" on the female dominated pool deck), as well as a poet, an artist, a concert cellist, historians, scientists, in fact every expert you would need for the perfect Trivial Pursuits team. It is an intellectual Utopia where the only mundane chore to disturb your stimulating quotidian amusement is the laundry.

There is a US$5 per small bag service available but it gets very overburdened particularly after a five-day sojourn in a hot, grubby country. I am trying to save time and currency by hand washing most items and luckily I have a dryer – my humid, windswept balcony – although salt stains and tropical downpours both conspire to ruin my efforts.

There is a kind of caste system on board in relation to room allocation with senior faculty getting all the plum accommodation on the top deck with sun lounger sized balconies and bigger rooms. Less senior faculty have reasonably large rooms, with baths as well as showers, and a few lucky ones have small balconies. As you descend the ladder of rank, so do your room, bed dimensions and toilet facilities deteriorate right down into the bowels of the ship where pullout wall bunks prevail. However, the crew are reputed to have their own gym and social bar down there and, perversely, we're all dying to see it.

Time to plan my evening, now, with the help of the daily memo. Shall it be a Dr. Phil-style self-help lecture on " ", or perhaps hide in my room and watch Ralph Fiennes in "The Constant Gardener", or maybe fall asleep in the "Web Page Team Interest Meeting"? No, I think I'll join the other boozers in our local "pub" for the 50th "knees up" and pick brains for the best travel options in Cape Town.

Part Three

From the tomato soup onwards, our arrival in South Africa was a total joy! Frustrating fog problems delayed disembarkation and completely concealed Table Mountain, but such irritations were assuaged by three bowls of soup on board and then we were let loose in the amazing Victoria and Alfred (her son, not hubby Albert) Waterfront in Cape Town. As the tablecloth cloud covering lifted, the prehistoric mountain backdrop did not look real even when we were standing in it! The city is set on the foothills sloping verdantly down from the famous mountain and on flat coastlands reclaimed from the sea in Table Bay.

It is a beautiful metropolis, with curved sandy beaches extending right and left, layers of topnotch housing dotting the hillsides, lush foliage interspersing the development, and ultra modern buildings complimenting the traditional European architecture. The rich Dutch and English colonial heritage here is encapsulated in its food (high tea and boerewors sausage), buildings, street names, left hand-side driving, and the distinctive Afrikaans language. Everyone speaks English too, making travelling, talking and transactions very easy.

Seafood and game restaurants proliferate in the Waterfront, giving gastronomic gratification to the thousands of European, American and Asian tourists thronging the lively, shoppers' paradise. Craft markets selling high quality African memorabilia compete with modern malls for your "rand" but prices are very cheap compared to Canada due to the favourable exchange rates. Sea lions and seals fight for safe havens to bask in the sunshine on wooden docks well out of shark reach.

Service quality and food standards are very high at the Waterfront whether you are eating in a pub, street café or silver-service restaurant. Portions are large, with prawns looking more like lobsters and skewers of kudu, impala, ostrich and crocodile meat requiring massive oval platters (and similarly massive stomachs) to accommodate. Throughout there is an air of prosperity, hedonism, materialism and mass tourism with very little poverty visible. Everywhere is the tribal sound of Africa, with drums, choral singing, and amazing ethereal music emanating from electric guitars made out of olive oil cans.

Nelson Mandela's island of 18 year incarceration is a half-hour catamaran ride away from the Waterfront, accompanied by cavorting dolphins and penguins. We were very fortunate to have South African civil rights activist and world-renowned poet, Dennis Brutus on our Semester At Sea trip to the infamous prison. Dennis was an inter-port lecturer on board between Salvador and Cape Town and had been outlining to students and faculty the history and politics that had led to his own imprisonment on Robben Island with Mandela during the sixties.

Great, great grandson to John Dunn (a famous Scottish/Zulu chief), Dan Dunn was our tour guide to the island and he immediately recognized Dennis from television interviews. "You're not as tall as you look on TV!" was his first comment. This lead to much animated conversation about the prison, apartheid and Mandela, and then Dennis was asked to help in directing us all around the island, along with the other former prisoners and prison guards.

Encapsulating Mbuntu philosophy, Modise Phekonyane, a former inmate of Robben Island, has totally forgiven his Afrikaaner jailors. Although they "took away his youth," tortured him, incarcerated him for more than six years for his political beliefs and love of his people, and tried to sap his spirit, he preaches forgiveness, freedom and friendship with all races. He came to Robben Island full of hatred and bellicosity at the age of 17. His

contact with Mandela supporters in the prison diverted his attitudes towards peace and reconciliation in order to achieve prosperity for all. He said "If I turned a gun on the white people, then that would have made me just as bad as them. If I forgive my captors, then I am helping humanity move forward to create a great South Africa for us all. Just look at the mess created by fighting throughout the rest of Africa – in Angola, in Zimbabwe, in Rwanda."

What was so inspirational about Modise was his complete lack of bitterness about his terrible experiences in the overcrowded, hand-quarried jail. He blames the system of apartheid rather than its implementers, saying that the guards were for the most part just following orders and would have been just as likely to be abused if they had refused to carry them out. During our tour of the bleak prison, he told us lyrically and emotionally about life with no rights, no personal space, no privacy and no freedom. It was actually Valentine's Day but for him this Hallmark holiday holds no lovey-dovey sentiment, but rather terror, as it was the day in 1978 that he was thrown onto the former leper island for a five-year sentence.

To chat with Modise, listen to his stories and poems, shake his hand and have photos with him were the highlights of our trip. He is about to go to Cambridge, UK, and then onto America for Master's studies in Theology and fully intends to return to politics in the new South Africa afterwards. We bought his book of poems, entitled "*Out of the Storm of Robben Island: My Freedom, My Passion*" which he signed and dedicated for us. I fully believe we met a future President!

Now, don't get me wrong, there is dreadful poverty, hunger, disease, terrible infant and adult mortality, political dissatisfaction, crime and still some racism in South Africa. However, there is an increasing black and coloured middle-class emerging as well as government, private and charitable efforts to house the dispossessed and clean up the shack-squalor of the townships. These shantytowns, made of bits of wood, plastic and corrugated iron, dot the highway to the airport and share views incongruously with the pristine, coastal suburbia of beautiful Hout Bay. But they do have electricity now and their very existence underlines the peoples' right to live in Cape Town, a privilege disallowed under apartheid before the ANC became government in 1994.

Townships are now waking up to tourism and encouraging foreigners to visit their bars and restaurants (with guides) in order to raise their profile and finances. Semester At Sea students frightened their trip leader by overindulging in a variety of wines and cocktails made readily available in the restaurants on their township tour, ostensibly promoted as a cultural music visit. On a tourist bus trip around the Peninsula, where the city's rural suburbs sprawl to the edges of the Cape of Good Hope National Park, you can disembark at various, very different bus stops. First the lush botanical gardens, then an amusing bird and monkey sanctuary, next a distressing, poverty-ridden township, followed up by ice creams at a prosperous seaside community reminiscent of Hove seafront on England's south coast.

We didn't make it to the famous wine country which surrounds Cape Town to the North and East but fellow passengers raved about it – albeit with bloodshot eyes and obvious headaches! The supermarkets, wine stores, cafes and restaurants are full of choice South African wines at such reasonable prices. It makes a US$8 bottle of Simonsvlei Rose at lunch an absolute necessity!

After our amazing safari into the Timbavati Private Game Reserve, I am a confirmed "Hunter Gatherer". In Calgary, the nearest thing to bagging my own prey is finely tuned bargain hunting at the Boxing Day Sales. During our four-day stay, my city-dulled senses all came to life, making eyesight keener, hearing improved, and smell heightened by the raw intensity of the bush and the thrill of the chase. In open-top, eight seater land rovers, we sped through the lush, green forest seeking out both prey and predator, alternating roller coaster speed with intense silence in order to watch these beautiful animals at work and leisure. In safari-talk, we saw four of the Big Five, (lion, leopard, elephant and buffalo), missing only the elusive rhinos who hang out in the Kruger National Park bordering our private reserve and sometimes pop in for water. But, in Louise-talk, we saw the most beautiful creature in the world – the leopard.

Over three 5:00 a.m. start days we had much intimate contact with several different leopards. With my new perspicacity, I was able to differentiate between two females just by the size of their spots! Mind you, after 25 years of designing and making clothing from leopard-print fabrics, I did have some advance training. The most dramatic, heart-pounding encounter was a nighttime pursuit through dense forest, thorny boughs whacking us in the face, giant bugs committing oral and nasal suicide, until we spotted the spots in a flash lit clearing. This was a sleekly powerful male who was more interested in dinner than us. Thank goodness there were so many impala around! Their proliferation throughout the veldt plus the white furry "M" emblazoned on their buttocks has led locals to call them MacDonalds.

Drunk on the endorphin rush (and the Cape wine served in a clearing at sunset on leopard-print tablecloths), we were then treated to a bloody hyena feast. No polite table manners here, the huge hyena just scoffed and scouted nervously for competitors, smiling gruesomely into our spotlights. It, too, was beautiful despite its odd stature and blood-soaked jaws. Its soft, fluffy fur looked freshly shampooed and eminently pet-able.

The poor female leopard that had actually brought down the impala was panting nearby, recovering from her exertion and distraught at having her kill purloined before she could drag it up a tree. That evening we also spotted huge, male elephants, a hippo blundering across a clearing, and a herd of wildebeest hanging out with their symbiotic buddies, the nervous zebra. We also spied on a pride of six female lions, reclining in the bush, aware of us, but not bothered enough to disturb their respite from hunting. Next day we came across a different female leopard, which had successfully treed her plat du jour and was bloatedly sleeping it off, legs akimbo in her branchy hammock.

A safari is comparable to living in an Animal Planet TV show with hushed voices and semi-controlled excitement. Every insect, plant, bird and creature becomes important; every footprint could be the key to a majestic presence nearby. A team of guide/driver and spotter, who perches bait-like on a precarious seat over the front headlight, mans each vehicle. Their field-craft and knowledge of flora and fauna is impressively extensive imparting to us city slickers the best hands-on biology, botany and ornithology lesson imaginable.

Dawn and twilight motorized safaris are complimented by a morning bush walk, which concentrates on small but exotic fare including gigantic cartoon-coloured golden orb spiders in guy-rope webs, industrious dung beetles, termites adding to their 50 year-old ancestral piles, and brilliant birds, such as the pastel pink and blue European Rollers. A kaleidoscopic treat was provided by a swarm of Carmine Bee-Eaters who dived and swooped around us, banqueting mid-air on frantic crickets and grasshoppers. The speed, proximity and the multi-coloured plumage of this aerial ballet made us feel like we were inside a magical flickering rainbow.

On our journey into the reserve we saw a gossiping group of baboons, a guilty-looking jackal with only one ear, several enormous stately giraffes feasting on top branch foliage, three treetop funereal vultures, and two skittish zebras. After a three-hour flight from Cape Town to Nelspruit's brand new thatched-roof airport via Johannesburg, plus a three-hour bumpy drive, these first sightings managed to reanimate even the sleepiest student on our bus.

Our "camp" called Motswari, is utterly breathtaking in its tribal luxury and four-star service. Personally I would have called it five stars but manager, Kathy Bergs, insisted that there's an even higher-class operation in Timbavati, which lays claim to that ultimate designation. Imagine fifteen idyllic thatched-roof round bungalows, with shady porches surrounded by acacia trees. These honeymoon cottages encircle the central breakfast house, "boma" (alfresco dining area) and lunch lounge with its indoor/outdoor trees and resident Ververt Monkeys. With gleeful speed and agility, they rivaled Brazilian street thieves in their daring banana thefts from our buffets.

When you're really exhausted from crack of dawn safaris, whole days spent outdoors, sweltering bush walks, enervating 30 degrees Celsius, and bestial gorging on barbecued kudu, ostrich and impala (us, not the leopards), what you really need is a great night's sleep in a fabulous bed. Our beds were SO comfortable that we actually found ourselves fantasizing about them during daydream moments. Each bungalow has two queen beds put together, festooned with white linen enveloped feather duvets and pillows, under vast swathes of sheer mosquito netting. It was like sleeping in a giant crib made of clouds.

Even mosquito repellent was provided in the huge bathrooms although we hardly saw one of the malaria-carrying monsters. The décor throughout was plushy tribal, with animal rugs, bark-covered lampshades, rattan, wood and stone furnishings and obligatory, fierce native masks. Wireless Internet was a surprising addition in the Gallery that also featured evocative watercolours of the Motswari animals painted by Marion Geiger, the reserve owner's daughter. Free Internet to both students and faculty was like free bananas to the Ververt Monkeys.

Our convivial guides, spotters and hosts, with whom we spent almost the whole day and evening, enhanced our entire safari experience. Everyone ate together around the campfire in the alfresco "boma" or at the surprise bush barbecue, where tables were immaculately laid in a secret clearing with a four-course Bacchanalian feast, with only flickering firelight to compete with nature's starry overhead streetlights.

In the afternoons, my son, Fergus shared the rock-lined swimming pool with a tiny terrapin and, one day, was almost lunch for a fiery-tempered warthog. After staring him out, the comical bush boor finally relented, seemingly deciding to remain vegetarian after all. Simon's shorts were also a potential predator's snack. Hyenas scavenge anything at night in the Motswari compound and left-out laundry is apparently irresistible.

Heart-stopping nocturnal encounters in our luxurious camp are not uncommon, so we have to be escorted to and from our huts by our trusted trackers. Indeed, we even had to be careful which bugs we inadvertently ingested as swallowing the pretty Monarch Butterflies has been known to cause heart failure.

Butterflies, in matching colour collusion with flowers, were amazingly prolific all over the picturesque South African countryside. On our journey back to Cape Town, we noticed that the rural villages and towns between Nelspruit and Kruger were also teeming with pedestrian native people, some festooned in colourful tribal cloths and headdresses, others sporting western fashions topped with umbrellas. Everyone was wearing shoes, except a few very young kids, and, judging by the level of commerce going on by the roadside with its cafes and strip malls, I don't think bartering bras for goods would have been quite the done thing. One square brick restaurant was optimistically called "Young Executive Café" and another, amusingly, "Chic and Chip." Houses were also brick with curtained glass windows plus moderate smallholdings for subsistence farming.

The highlight of our last day in Cape Town was having miniature Squirrel Monkeys climb all over us, attempting to open Velcro, buttons and zips and steal anything they could find with their tiny fingers and perfect nails. This visit to The World of Birds, at Hout Bay, was the boys' choice – personally, I would have preferred the township stop-off where you can visit the Kroonendal Project, a job creation charity set up to teach unemployed township women to hand paint teabags and curios.

The small part of South Africa we saw was enough to make us all want to come back here whether as a tourist again or as a resident sometime in the future. The climate is refreshingly Mediterranean; the people are open, friendly, optimistic and industrious; the scenery encompasses all extremes of nature's beauty; and, in this model of peaceful rapprochement, there would be myriad opportunities to make a positive impact towards helping turn poverty to prosperity.

For now, though, we are just four days away from the beachy paradise of Mauritius. I'm frantically spotting up on my French, cutting back the calories for bikini-readiness as well as making space for some fabulous cuisine francaise, and planning a trip to the Black River Gorges National Park in search of yet more monkeys.

Louise Hudson is currently sailing with family around the world.

Along the Quetzal Highway

Tom Gehrels

I was invited on a trip along the Quetzal Highway, the backpacker's route that runs from Guatemala to the Mayan Riviera. I tried to come with no expectations. Beaches aren't my kind of thing and seeing the trip ended in Cancun, I had a feeling the focus was going to be sand and sea. I was to be pleasantly surprised!

The tour started in Antigua, a lovely town with cobblestone streets and colourful houses, a lively town square and lots of boutiques, cafes and restaurants. UNESCO protects the town; many ex-pats have made it their home, while indigenous people sell their textiles in the many plazas. We didn't spend too much time here due to time constraints, so I ended up spending the afternoon running around taking in the town, the shops and the sights.

Next stop on the itinerary was Rio Dulce, where we stayed in bungalows perched on stilts above the lake. A pool and a bar were the centerpieces of the resort and after a long night at the bar we were glad to relax the next morning on board a boat. We made a trip down river through a beautiful canyon and ended up in Livingstone, a laid-back English – speaking town on the coast and home to the Carib people. Lunch was had here and we spent a few hours checking out the shops and a friend even had his hair plaited. We travelled back by boat to Rio Dulce and on by bus to the quaint town of Flores, situated on an island in Lake Pent Itza. From here we visited the most imposing of the Mayan cities, Tikal. Set among a series of jungle trails the town boasts awe-inspiring plazas, temples and pyramids while the sound of howler monkeys can be heard in the background. Here we engaged in our first vigorous activity, climbing the steep stairs leading to the top of the pyramids. The heat made it quite hard to climb up but once on top the views made it more than worth our while.

An unpaved road took us through the backcountry of northern Guatemala and into Belize to the frontier town of San Ignacio. Some of our group relaxed by the pool, while others hiked in the rain forest or went caving. The latter is an especially unique experience. We waded through water up to our chests through an underground river and after a few hours of worming through tunnels and narrow passages we reached a ledge where we found ancient Maya pottery and skeletons (from human sacrifices) strewn out on the cave floor.

What struck me in Belize is the pride the people take in preserving their natural heritage and nowhere was this more evident than on the reef off Caye Caulker, an island along the Caribbean Coast. The snorkeling here is some of the best in the world, with schools of multi-coloured fish surrounding you as you swim between coral formations. If you are lucky you may even see some bigger marine life such as a barracudas or rare manatees.

Caye Caulker itself is a laid-back strip of houses, restaurants and pubs and a perfect place to swap warrior stories from all the different optional trips one can choose from. Diving, manatee, excursions, shark feeding, snorkeling and shopping are but some of the activities on offer. Some of us went sport fishing and as true sports-fishermen do, they drank more beer and caught more sun than they caught fish.

Onwards by speedboat and bus we went to the resort of Playa del Carmen in Mexico. It is at the heart of the Mayan Riviera that I had dreaded so much and once again I was pleasantly surprised. The people were lovely and while there is a lot of development, the town doesn't crowd in on the beach and there aren't any high-rises close to the beach. There was time to relax on the beautiful white sandy beach or swim in the azure blue water, while some of us took an optional afternoon trip to the beachside Mayan ruins of Tulum.

Our trip ended in Cancun, a bustling tourist resort with long sweeping beaches from where several small quaint islands can be easily visited by boat. Some of us stayed a few days before returning home, while others continued on the next leg of the trip, down through Southern Mexico and the Guatemalan highlands back to Antigua.

This trip is a great quick breakaway from the everyday and highly recommended for travellers who like the freedom to choose their activities and who want to see and do a lot of different things in a short period of time.

Tom Gehrels is with Trek Escape, Toronto, Ontario.

I hear & I forget
I see & I remember
I experience & I understand.

Confucius

California Freeway

Mike Dolphin

I have been living in California for just over four years now, attending Stanford University for a PHD in aeronautical engineering. I grew up in Vancouver, and have lived also in London, Ontario, (attending University of Western Ontario).

Stanford is located in the heart of Silicon Valley, halfway between San Jose and San Francisco. Everyone owns a car, and driving is the norm. The Bay Area is essentially a giant suburb of over seven million people (and thus seven million cars), with no central hub city to speak of. With all these people driving around all the time, you'd think the road signs would make more sense.

The first fundamental problem in California is the direction commonly referred to as "north". What is north? Perhaps I was privileged to grow up in a city where north was simply the direction of the mountains (the North Shore), at least while in Vancouver. I have a good sense of direction in any city though, perhaps in the woods and mountains of my hometown. I'm one of these people who can answer in an instant which way is north. Ironically, that only makes matters worse around here.

Californians believe that their coastline runs north-south, which is true, kind of. Actually, the coast runs northwest-southeast for the most part, and admittedly, this is cumbersome to say. The next logical step then is that if you are inland, the coast is to the west. Again, often true, but frequently not. For example, ask a resident of Thousand Oaks, California, (this is near Los Angeles, I am familiar with Thousand Oaks, as this is where my fiancée grew up), which direction they would need to travel to get to Malibu, which is on the coast. They will invariable say: "West". Obviously. The coast is to the west. Except that if you look on a map, Malibu is due southeast from Thousand Oaks. Only off by 125 degrees… Ouch! To confuse the matter, the 101, which is the main freeway through Thousand Oaks, is labeled "101 North" and "101 South." It runs due east west. Confused yet? To put it mildly, the Freeway signage here in California is a gross oversimplification of reality. The freeways twist and turn up and down the coast, generally going from one place to another in relative straight lines, but often changing cardinal directions along the way. However, the freeway signage is restricted to four directions: north, south, east, or west; and never changes, regardless of the actual direction of the road at the time.

Here are a few skill-testing questions for those of you navigating at home.
Q: You are driving in the East Bay near Hayward on the "238 South." You are approaching a freeway, which runs perpendicular to the one you are on. The sign for the interchange on-ramp says "580 West". Which direction will this take you, left or right?
A: Easy one, right? Nope…. Left. In this bizarre case, the "238 South" goes due east. Where it meets with the 580, the "580 West" goes pretty much north. The confusing result is that you must turn LEFT to go from "south" to "west."

Q: You are driving onto a freeway in Berkeley, and the on-ramp sign says "80 West." Then you notice another sign that says "580 East" for the same on-ramp. Quick, you only have a minute to figure this out before you miss the turn, which direction are you headed?
A: South. Yes, there is an on-ramp in Berkeley, which is labelled both "80 West and 580 East," and it actually goes due south. Oh, and it will eventually become the "880 South," but it doesn't tell you that at the on-ramp.

Q: You are "Southbound" on the 280 (a major freeway in San Jose), and you wish to be going "Northbound" on the 680, which direction do you turn?
A: Go straight. This is a trick question. The 280 become the 680 in the middle of San Jose. At that point the freeway is heading east west, although the 280 is still called "South" and the "680 North." (Actually, the end of the "280 South" is actually heading almost exactly north-east!). Ok, so that's just the freeway, right? I mean it's somewhat understandable to pick a system and be consistent, even if it's wrong in places. The "580 West" goes west sometimes, and north sometimes, but it's always the "580 West." Okay, I suppose I can get used to that. But you might as well call it the "580 up" and "580 down"…. it would make just as much sense. Or maybe call it "580 in" and "580 out". I could see an equally strong argument for these. Or how about "580 there" and "580 back", I'm sure we could come up with lots of options.

The East/North roading however, is pervasive. If you are in San Jose and ask someone which direction it is to Santa Cruz, 99 out of 100 people will say "west. The coast is west". In reality, Santa Cruz is due south of San Jose. Or people at Stanford will say "go east on Palm Drive". Ummm, Palm Drive is pretty darn "north" last I looked on a map. In fact, it is interesting to note that Palm Drive was laid down pointing exactly at the magnetic north direction of the time, approximately 18 degrees east of north. (Magnetic north has moved slightly since then.) East on Palm Drive will take you right into the curb. I don't know if there are any crazy examples similar to this in Canada (there probably are), but it seems to me that if a road which is generally west-bound becomes due north-bound, the best solution would be to do one of these things: rename the road "North" for sections the road goes north; rename the road a different number to avoid confusion; use more than just four directions (for example, NE, NW, SE, SW).

My other local traffic favorite is that on most major residential streets in Palo Alto, they seem to insist on posting "one-way" signs on BOTH sides of the street at the intersections. This is not only confusing, but it's completely wrong. The road is CLEARLY a two-way street. It just so happens to be bigger than others. Often, these streets will have a barrier running down the middle, but not always. Even still, they don't call it "EL Camino Real North," or "El Camino Real westbound" or anything like that… it's simply called "El Camino Real"… and thus does not warrant a "one-way" sign on either side. For example, the intersection of El Camino Real and Page Mill has four "one-way" signs… pointing in all four possible directions. Thanks!! Here I am approaching a major intersection, and as I near it, I see a one-way arrow pointing to the right. Da!, I was going to turn left, but wait, there is a left turn lane? Huh? Oh, wait, now I see ANOTHER one-way sign pointing left. Um. Yah. All I ask is why?

I should note that although Palo Alto uses this dumb double one-way sign concept, not every municipality decides what the signage should be. With people driving all over the Bay Area, and all over the State for that matter, shouldn't the signage be standardized across the whole state? Not determined by what the local "city " council decides? I put city in quotes, as each "city" is really an extension of the previous one, that is, one suburb bleeding into another. Albeit, each with their own signage and parking restrictions. (Which is why I was shocked to discover that I was not allowed to park on the street in front of my own house in Menlo Park between the hours of 2:00 a.m. and 4:00 a.m., everyday.)

All things considered, once you adjust, the signage is not terrible, except for the off-ramp signs. These signs are usually placed AFTER the off-ramp splits off the main road. Thus, unless you have very good vision, the sign is effectively useless until it's too late to change lanes. Now that is really annoying! However, once you live here for a while and know where you are going, who needs signs, right? Apparently, I still do. I still scratch my head every time I get off the 280 to go in the direction I intuitively know is north. The answer to this, is to go "Page Mill East." Luckily, there is also a sign that says "Stanford, this way". Thank god. I would never have found it otherwise.

Mike Dolphin is studying at Stanford. He is originally from Vancouver, British Columbia.

The traveller was active;
he went strenuously in search
of people, of adventure, of
experience. The tourist is passive;
he expects interesting things to
happen to him.
He goes "sight-seeing."

Daniel J. Boorstin

Hawaiian Greetings

“Ahoha ka ka hiaka’ - Morning Greeting
‘Aloha auinala’ - Afternoon Greeting
‘Aloha ahiahi’ – Evening Greeting
‘Aloha nui loa’ – Til we meet again or see you soon
‘Matama Pono’ – Take good care
‘Mahalo nui loa’ – Thank you very much!

Non-stop Solo Circumnavigation Victoria To Victoria

Tony Gooch

The idea for such a trip came up while Coryn and I were returning from a summer's sailing in Spitsbergen in 2001. We prefer sailing in the high latitudes, and the wilderness of the Southern Ocean holds a special attraction. I wanted to get down there one last time while I was still fit and healthy enough to handle hard sailing and enjoy it all. Such a trip had never been done from the west coast of North America (or, I think, from the east coast), so I decided to make the attempt from our homeport of Victoria, British Columbia. The first step was to get Taonui from England to Victoria. An 'old' sailing friend from Toronto helped me sail her across to Boston in June/July 2002, where she was trucked to Seattle and brought over to Victoria in mid-August.

A non-stop circumnavigation from the west coast of North America poses a different set of navigational problems than those encountered on a similar trip starting from Europe. The best time to leave Victoria is the end of September in order to avoid the hurricane season off Central America and arrive at Cape Horn at the start of the southern summer, then spend the months from December to February circuiting the Southern Ocean. Unfortunately this means returning north up the South Pacific during the cyclone season. A trip starting from Europe does not have this problem, as there are no cyclones in the South Atlantic. Starting from Europe allows for a gradual introduction into the Southern Ocean after sailing south of Africa, with the final hurdle of Cape Horn at the end of the circuit of the Southern Ocean. At the end of the first leg I would be sailing directly to Cape Horn, with the dangerous south coast of Chile as a rapidly approaching lee shore. A further consideration is that the last leg up the Pacific involves threading a course through the various island chains and dealing with strong ocean currents, neither of which are a factor in the return to Europe up the South and North Atlantic.

I estimated that the trip would be about 27,000 miles and would take between 200-210 days at an average daily run of 135 miles. If I left at the end of September I should be home before our wedding anniversary on April 29th, 2003.

The attempt was to be conducted under the auspices of the UK based World Speed Sailing Records Council, which keeps track of offshore sailing records. WSSRC regulations stipulate that I could receive no assistance, only use the motor to charge the batteries and could only anchor to effect repairs.

I crossed the starting line between our house and the lighthouse on Trial Island, just before noon on September 28th, 2002. I had on board enough food for eight months and 210 cans of beer… a ration of one a day. The first 60 miles was a stiff beat down the Strait of Juan de Fuca and out to the open sea. I had expected strong north and northwestern winds down the west coast of the US, but in any event the winds were generally light all the way to the ITCZ (Doldrums) at around 10 degrees north. The average daily run for the first three weeks was a disappointing 125 miles. One dark night while running with poled out #1 genoa and drifter, a sharp increase in wind blew out the drifter, which had seen many miles and too much sun and was well beyond its "Use by Date." It was quite a struggle getting it down and bagged for a decent burial when I got back to Victoria.

In the rainy squalls of the ITCZ it was easy to top up the water tanks, which gave me a revised ration of 4.5 litres a day for the expected 25 weeks needed to get back to Victoria. I picked up the SE trades at 4N and sailed close hauled down longitude 117W. Gradually the wind backed to the east and then the northeast, as I was lifted down the western side of the South Pacific High. At 35S the wind had backed all the way to the northwest, and I was starting to feel the force of the fronts spinning north from the lows as they circled Antarctica at 50-55 degrees south. At 42S, I saw a wandering albatross, a sure sign I was back in the Southern Ocean. On November 18th I was at 47S with 1,080 miles to go to Cape Horn when the barometre dropped from 1008 to 990 in 15 hours. The Chilean weather fax showed a large-area low moving northeast that would give strong southeast and then southerly winds for some time. I decided to lie to the drogue and let it all go by rather than sail close-hauled into such strong winds.

While I was getting the drogue out of the cockpit locker, Taonui was knocked flat by a large wave. I had time to turn my back to the wave and drape myself in front of the open cockpit locker, but still gallons of water went into the stern compartment. Fortunately, the companionway was shut, but the dorade over the navigation desk was open. Later I found about an inch of water in the containers at the back of the navigation desk, and the radar tube, which is directly below the dorade, took a dunking. My heart sank when I pressed the display buttons and water squirted out, so I kept pressing until that stopped. Later I took the display unit apart, and thankfully found it quite dry inside. It did stop working later in the trip, but that was for another reason.

The barometre stayed at 990 for 12 hours and the wind blew a steady 45-50 knots with gusts to 55. I lay to the drogue for 39 hours until the barometre rose to 1000 and the wind slowly shifted to the southwest. The drogue, which is deployed from the stern, is a 250 foot long one inch line on which are mounted a total of 120 nine inch wide cones. It slows the boat down to about $1\frac{1}{2}$ knots and is a great defense system. Unfortunately the retrieval line snagged while it was running out, so I had to wait until the wind dropped to about 20 knots before I could pull it all back on board.

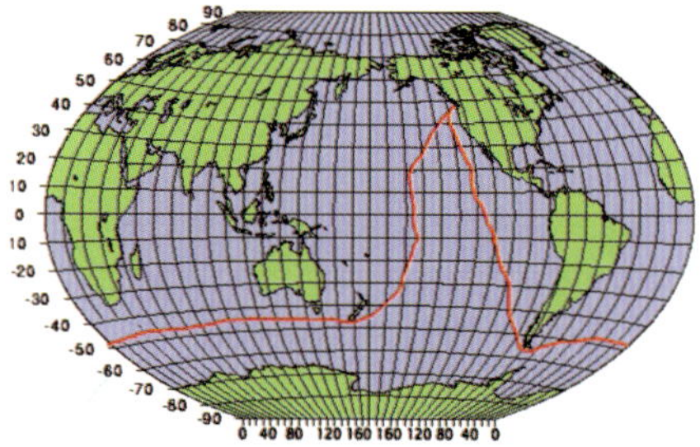

For the next week, the wind followed its usual pattern of shifts, NW, SW, W, NW and stayed between 25-35 knots giving some fine sailing as I ran off the last 900 miles to Cape Horn. I had planned to pass about 90 miles south of the Horn to avoid the shallow bank that extends 70 miles south of Cape Horn. Here the depth rises from 13,000 feet to 300 feet in only 20 miles. In bad weather this bank produces horrific seas. As it happened, on November 26th, the winds were only 20-25 knots from the west so I could sail over the bank and close along the shore of Cape Horn Island. The word 'awesome' is totally appropriate for this place, known to the sailors of old as "The Uttermost South" (until they found Antarctica). I was lucky enough to see it in bright sunshine, with smooth seas and a dusting of new snow on the hills of the nearby islands. Taonui had logged 7,457 miles since leaving Victoria 58 days previously. I felt a strong sense of gratitude, accomplishment and relief to have made it around Cape Horn, and although there were still many miles to go it felt as if I was halfway there. No longer was I sailing south and away from Victoria, now I was sailing for home.

Just before leaving Victoria, I installed "sail mail" on the boat. The components are a laptop computer, a Pactor modem and a SSB or Ham radio. The service costs $200 a year for an average daily transmission time of ten minutes, which was ample for Coryn and I to exchange emails every day. Coryn sent me weather information and lots of loving support and I sent daily reports which Coryn edited and posted on our website, which we created to allow friends to follow the trip. Over the course of the trip, the site became known and progressively developed a surprisingly large daily following.

Coryn relayed reports from the Argentinean Navy warning of considerable ice in the South Atlantic. Sure enough, on November 29th I saw what was to be the first of many icebergs. A 20 x 50 mile piece of the Weddell Ice Sheet had broken off two years ago and the remnants were strewn over the South Atlantic, some as far north as Montevideo. I increased the radar watch to 20 sweeps every ten minutes. The radar seemed to see the large icebergs quite well, but did not always see the growlers, though fortunately there didn't seem to be too many of these. Sometimes the icebergs were so numerous that there were always one or more on the screen. This cut down on sleep time as I couldn't be sure that the radar alarm was from a berg that I had seen or a new one that the radar had just detected. A week later I came upon a monster "table" berg, perfectly flat on top and about 200 feet high. The two faces that I could see measured six miles and four miles on the radar. It was a bright sunny day and I was able to sail quite close to this majestic piece of natural wonder. Since leaving Cape Horn at 56S, I had eased north and by December 8th I was at 47S and appeared to have reached the northern limit of the icebergs, or so I hoped.

I had planned to sail the Southern Ocean at around 44S. The lows seemed to be tracking along between 50 and 55S, so I stayed along latitude 47S for the next seven weeks, using the weather faxes from Chile, South Africa and then Australia to monitor the progress of the lows. Additionally, Coryn emailed daily predictions from Meteo France and the NOAA weather forecast models that cover all oceans. Sailing along 47S as opposed to 44S cut about 1,500 miles from the distance to be sailed. Most of the time the wind oscillated from NW to SW, usually Force 6 or 7 occasionally Force 8 or 9. Sometimes it dropped to Force 3 or 4. On one light wind day I had a visit from a Fin whale. Typically 85 feet long, the Fin is the second largest of all the whales, and I guessed this one to be 65 feet. For ten minutes he

swam close to Taonui, surfacing off her stern and swimming alongside, then disappearing and sliding under Taonui to resurface on the other side and repeat the performance. I just stared at him in dry-mouthed wonder, as I knew that with one flick of his tail this trip could have a different ending. However, it was clear that he had perfect control and posed no threat to Taonui. The Fin whale is unique amongst whales having a large white patch on one side of his head while the other is totally brown.

One of the pleasures of the Southern Ocean is the prolific bird life. I spent hours marvelling at the effortless gliding flight of the albatross, particularly the Wandering Albatross. The sea is rich in food and supports huge numbers of petrels, prions and shearwaters most of whom nest on the sub-Antarctica islands. It is one of the world's great wilderness areas. On a sad note, the population of albatross is in precipitous decline due to the irresponsible fishing practices of long-liners. Unless the lines are screened in some manner the albatross go after the baited hooks as they are being set and are dragged underwater as the long line sinks. The impact has been particularly devastating for the wandering albatross as they mate for life and raise only one chick every second year.

On December 25th I was at 47S 41E, just east of Marion Island and about halfway across the Indian Ocean. The seas were so light that I was becalmed for four hours. Santa, in cahoots with Coryn had left a bag of presents and loving thoughts. Christmas dinner was special. Fresh free range Marion Island turkey, so fresh I had to remove the last of the tin coloured feathers before stuffing it, asparagus, plum pudding and cream, and a fine bottle of wine with toasts to Coryn, friends and family and to Taonui.

January 7th was definitely not a good day. In the morning, I rushed on deck in response to a loud bang and found the wind generator hanging on two wires. These soon snapped and the generator fell into the sea. Weeks of strong winds must have caused metal fatigue in the support pole and it snapped off at the base in a particularly strong squall. One way to reduce weight aloft!! (Three weeks before I arrived back in Victoria the radar stopped working. When I got home I found there was a large hole in the back of the radar dome that was no doubt caused by the falling wind generator.) Later that same day, as the wind gusted to 50 knots, I decided to change down to just the staysail and in the process the #2 genoa became wrapped around its foil and before I could unwrap it the top half was blown to shreds. Fortunately I had a spare, slightly smaller jib, which I was later able to hoist in its place.

As I approached Australia, the lows were starting to move further north so I eased up to 45S for a week or so to keep some distance from the stronger winds at the center of the lows. On January 15th, Cape Leeuwin on the southwestern corner of Australia was about due north and I rode the northwest winds back down to 48S making a course for Snares Island at the south end of New Zealand. The passage south of Snares Island is not a good place to be in bad weather. I was lucky and passed over the shallow banks with a Force 6 to 7 from the southwest. The next day the New Zealand forecast was for 60+ knots at Snares! The distance sailed from Cape Horn to Snares was 9,553 miles covered in 63 days at an average speed of 152 miles a day. Although the Southern Ocean holds a compelling fascination, it is hard work and I was ready to head north.

I decided to cut the corner and head directly for Victoria rather than follow the planned route of sailing 2,000 miles further east to 150W before turning north. By February 1st, I was up to 43S and beating into light NE winds to get around Chatham Island, which lies about

200 miles southeast of Wellington, NZ. I expected to be on a starboard tack for most of the next 7,000 miles to Victoria, so I shifted all moveable weight up to starboard. I set a course of 70 degrees True to a waypoint of 39N 150W, which I hoped would keep me clear of cyclones. A large area of high pressure extended east from the Tasmania Sea and for the next week or so I had some fine close-hauled sailing in brisk East to SE winds. On February 6th, I was at 36S 166W and starting to worry about a cyclone named 'Dovi' that had developed at 18S 163W. The wind had backed into the east and it was going to be a struggle to keep out of Dovi's way as it moved south down longitude 169W. By February 8th, I was 360 miles east of Dovi's centerline and 600 miles to the south. Troughs extending east from Dovi were generating east winds of 30 to 40 knots. With a double reefed main and staysail, I hung on to as much easting as possible as Dovi continued to move SSW and I sailed north. Finally on February 11, I was 600 miles east of the center of Dovi and the wind started to ease.

As is typical, behind the cyclone was a large area of no wind, and for the next five days I averaged only 81 miles a day. By February 15th, the weather systems had returned to normal. The South Pacific High had reestablished itself and for 28 hours Taonui romped along on a broad reach under full main and asymmetrical chute making seven knots. It was great for the morale to be moving again. At noon on February 15th, Mangaia, the southernmost of the Lower Cook Islands was a low grey lump in the ocean about 20 miles to starboard. Just being at the southern edge of French Polynesia gave a sense of progress. About 120 miles ahead were three more of the Cook Islands, which I will have to pass through the next day. After that, its clear sailing all the way to Hawaii... 2,500 miles away on a course of 10 degrees True.

Between 13S and 7N there were several bands of Convergence Zones. The sky was full of squalls with much rain and changeable winds requiring endless sail changes. The hot, muggy weather didn't help, nor did a strong current of 1 to 2 knots setting to the SW between 2S and 10N. On February 24th I somehow managed to catch the sacrificial strip of the # two Genoa on the radar reflector on the mast resulting in an 11 foot rip down the leach. I decided not to take the time to repair it as I expected strong close-wind sailing for most of the rest of the trip. I crossed the Equator on February 27th at 156W and by 10N I was out of the squally weather and into the clean, clear northeast winds. With no adverse current and Force 5 to 6 winds, the daily runs jumped from 135 miles to 155 miles a day.

I had hoped to pass east of the Hawaiian Islands, but ended up having to beat my way through the chain passing south of Maui in a light east wind and against a west going current. The trip from Hawaii to Victoria is usually a close reach against the NE winds until you are north of the center of the North Pacific High, at about 30 to 35 North, and then a run to the coast on the SW to West winds. Again I was lucky with the weather. The high had wandered off to sit over southern California and a continuous series of lows marched their way NE from Japan giving strong to gale force SW winds. I was able to sail a rhumb line course directly to Victoria. Having only one genoa I had to run wing and wing, rather than use poled out genoas. At 9:00 a.m. on March 19th a violent squall hit with 50+ knots of wind. Taonui rounded up and at the same time a larger wave hit her broadside. The boom dug into the water and the forward motion was such that the boom snapped at the vang. I was in the cockpit and watched it all happen as if in slow motion. A sickening sight and feeling. Luckily I had brought precut and drilled angle bar to fix a broken boom but I needed calmer weather to make the fix. With the squall the wind shifted into the NW and

stayed at 35 to 40 knots for two days. All this pounding was coming from a 989 low that was taking its own sweet time tracking along 50N and into the Gulf of Alaska.

I spent most of March 21st repairing the boom, as it was likely that I would find east winds in the Strait of Juan de Fuca. As it turned out, on March 24th I had a stiff beat up the Strait much of it against an ebbing spring tide. That night the wind died so I took down the sails and waited till dawn. Over 30 boats from the Yacht Club had come out to welcome me home, which was very heart warming after such a long time at sea. The last 10 miles were a real struggle, ending as a beat against a pale SE wind and an ebbing tide. I crossed the finish line off Trial Island at 11:00 a.m. Coryn jumped aboard and the local Coast Guard Auxiliary towed Taonui around to the Yacht Club. The trip had taken 177 days. The distance sailed was 24,340 miles, at an average speed of 137 miles a day.

I am very grateful to my wife, Coryn, for her role in this adventure. Aside from her help in preparing the boat and stocking her for the trip (including baking eight large fruit cakes), Coryn spent hours updating the website, handling newspaper, radio and TV interviews. Every night she checked weather reports, thought them through and emailed me a summary along with warm chat and encouragement. A large part of this trip's success and my pleasure, enjoyment and sanity is due to her efforts and support.

Tony Goochis a long distance sailor. www.taonui.com

Twenty years from now
you will be more disappointed
by the things you didn't do
than by the ones you did do.
So throw off the bowlines,
sail away from the safe harbour.
Catch the trade winds
in your sails.
Explore. Dream. Discover.

Mark Twain

Secrets of the Burnside River

Jeanne Davis

It was spring 1995. My new husband Gord and I decided to explore the Canadian Arctic. It would be a dream come true for both of us. We weren't seasoned Arctic travellers but one thing Gord knew was that we wouldn't see the Arctic from our truck. We would have to fly, float and hike to truly experience arctic wilderness. So after some research, we concluded that the Burnside River in the Northwest Territories was the right adventure for us.

Burnside Canyon

Since neither of us were experienced canoeists, we chose to float the river in a raft. Not that we were rafters either, but it looked less challenging than a canoe. A second-hand raft was found and cash in hand, Gord drove to Calgary to pick-up our new, second-hand 16 foot Metzger raft. The seller and the raft came highly recommended from expert rafters who assured us that the boat was in excellent condition. It was to be a perfect fit for our 300 kilometer river trip that would land us in the Arctic Ocean.

Twenty-four hours later, we were on the road to Yellowknife and the following morning we boarded a Twin Otter that flew us 500 kilometres north to Lake Kathawachaga. We unloaded all our gear that was to keep us floating, dry, warm and fed for the next 16 days. We checked one more time that everyone concerned knew where and when to pick us up – after all, the Arctic Ocean is a big place! We waved good-bye to our pilot and watched the plane disappear over the horizon.

Then an extraordinary feeling overcame us as we took in the grandiosity of the surroundings. Under an immense arctic sky, we gazed at rolling hills and tundra for as far the eyes could see. We were hundreds of miles away from everything with only a river to get us somewhere! We were excited and yet anxious. Large ice patches were scattered on the lake. The potential for an ice jam at the river mouth was real. We must get on the river soon or we could be landlocked for days!

Gord quickly laid out our new raft – a big orange flat boat. We began pumping up our new raft for the very first time. Our biggest worry was "What if there is a leak? No problem, we have patches!" But as hard as we pumped, the raft lay flat like a dead fish. The pump didn't work! After contemplating the situation for sometime, Gord announced that we would blow it up by mouth. New to the rafting world, we actually thought this was humanly possible! Lying flat on our tummies, we blew until we verged on hyperventilation. The raft remained flat. I was about to remark that it would take days to blow it up when Gord had the petroleum jelly flash. Maybe a bit of jelly would fix the pump. Our gear was spread everywhere as he searched for that tiny tube of jelly he was sure he had packed at the last minute. Yes, he found it – only a few small squirts left, but it was enough to fix the pump. The raft was successfully inflated and off we went challenging the ice flows on a beautiful sunny day in early July.

The next day and the day after, the sun shone brightly. We laughed that we had come to a place of no darkness with flashlights and spare batteries! It was probably on the third day that we began thinking of the petroleum jelly again. The arctic climate is very dry and windy. In all our planning, we had overlooked packing a lip balm. Our lips began to chap and then cracks developed. Soon we could no longer laugh and eventually, the faintish smile was too painful. The petroleum jelly became an obsession but the remaining few squirts were being saved for when we would have to deflate the raft for our last and longest portage and then re-inflate it for the final leg of the journey.

One thing about being in the middle of the Canadian tundra with a problem is that there is no use getting down. So even though we could no longer smile, we talked a lot about the beauty of the tundra and the hardiness of the plants and animals. We were awed by the total absence of signs of human presence. We were explorers from earlier times! With 24 hour daylight, we checked out every bend in the river, watching for the unusual boulder that turned into a muskox. We scouted the horizon for the caribou. We climbed cliffs to admire Peregrine falcon chicks. We lay in our tent admiring the Arctic wolf snooze by his den. We were determined to not let our lips spoil the adventure. We remained hopeful for a small miracle.

One day, we pulled up on an island in the middle of the river. We had floated by many such islands and each time decided to stop at the next one! Why we chose to explore this island, I don't know. As we strolled along the shore, we came to a very small bay, no more that a few feet wide. It was sheltered from the flow of the river. In the shallow water, a lip balm calmly floated. A cherry flavoured lip balm! Without a word, I picked it up, removed the cap, rubbed the tip on my shirtsleeve and proceeded to smear my lips up to my nose! Then I passed it to Gord. We didn't speak for a few moments. Truly, we were struck silent by a miracle on the Burnside River!

Arctic Exploration:
NW Passage to Sondre Stromfjord

Ned Ambrus

When I told people I was travelling to the Canadian Arctic and Greenland this summer, I got one of two responses: 1) Why on earth would you want to go there? or 2) Oh my gosh, how cool! Are you going to Beechy Island to see the Franklin Expedition gravesite? Where are you flying and what else are you going to see? I do feel the Arctic is a "love it or leave it" destination, but I promise that once you have been there, it will draw you back.

Was it the vastness of the Arctic, its sheer exquisiteness? Was it the way the sun glittered off the icebergs, or was it the deep blue hue of the ocean that entranced me? Or perhaps it was the endless sky, the crispness and clarity of the air, the stillness of nature, the beauty of the wildlife, or the lack of sunsets that had me hooked from the moment I arrived. Whatever it was, I fell deeply in love with our northern country, and I had one of the best holidays of my life that summer.

I travelled with a leading adventure company on a Category A2 icebreaker called the Akademik Ioffe. At a maximum capacity of 110 passengers plus staff and crew, I felt fortunate to only have 75 passengers on my trip. This allowed the staff to give us a little extra attention and answer any questions we had, from the types of birds we would see that day, whether we would need our wet weather gear for landings, or the mating habits of polar bears. We had an itinerary, which really was more of a suggestion than an absolute. Aaron, our expedition leader, always informed us at breakfast what our plans were for the day, and quite often they would change by lunch! Our daily seminars were extremely educational and informative, and I tried to sit in on as many as I could. However, the beauty of the Arctic kept drawing me outside, whether to see the icebergs from different angles, to whale or seal watch, or simply to gaze at the horizon. It was truly a photographer's paradise.

I chose to do the optional kayaking excursions. Graham, our guide, was very patient with the group, especially when he discovered (to his horror, I'm sure) that most of us lacked effective kayaking skills. But he persisted and by the end of the trip we were all shining stars! We even had a man overboard but thanks to Serge, our zodiac rescue boat driver; he was scooped from the water in no time. Being in the kayaks afforded us some truly magical experiences, such as checking out seals just a short distance from us (and being checked out in return), witnessing a glacier calving (a mass of ice breaking off from a large glacier) and exploring areas that even the zodiacs couldn't get into.

One of my fellow kayakers from B.C. had decided prior to the trip that she simply MUST swim in the Arctic Ocean, despite an average water temperature of two degrees Celsius and plenty of ice all around us. We are talking seriously cold here! And while we did wear dry suits and spray skirts, we always got a bit wet and were very thankful for the hot showers afterwards. Every trip out, Susan would persistently ask Graham, "Here? Can I swim here?" She always received a grunt of "no" in return. Quite often the water was too rough and the conditions too risky to allow her to jump in, much to her disappointment. However, the day finally arrived when the water was calm enough. After a fabulous 14km kayaking trip along the stunning cliffs of Refuge Harbour, Graham gave us the go-ahead to go for our long-awaited swim. Yes, US.

I figured if Susan could do it then, heck, so could the rest of us. While we initially scared the other zodiac drivers exploring the area, soon people started spilling out on deck to watch the silly kayakers go for a swim. I was quick enough to call "last in," figuring I would be submerged for the shortest amount of time that way. A quick, friendly argument ensued with one of my fellow kayakers, Craig from Australia, who commented that, "I should be a gentleman and let the ladies go first." My stubbornness won in the end. I watched everyone else do their swim and learned from their mistakes. So when my time came, I stood on the edge of the zodiac, plugged my nose and stepped into the icy waters. The cold was shocking and blasted my body, despite my attempts at preparing myself for it. I managed to do a 180 degree turn while still underwater and then promptly shot out of the water with such force that I came up over the edge of the zodiac to be grabbed by Graham and Serge. The entire boat exploded into laughter. Apparently my eyes were as big as saucers and Graham, a professional photographer, said he would have given anything to capture my exit from the water and the precious look on my face! Despite the dry suit, the shivers settled in very quickly. We made our way back to the boat, peeled off our gear, put away the kayaking equipment and raced for the showers. That was the hottest, longest and most satisfying shower I think I have ever had!

Our shore excursions included Beechy Island, final resting place for the ill-fated Franklin Expedition of 1845-46. Sir John Franklin was a British sea captain who died on an expedition in search of the Northwest Passage. He set off with a crew of 128 men in two ships, HMS Erebus and HMS Terror. They never returned. The gravesite is eerie and the artifacts are surprisingly intact, but the location of "Terror Bay" still gives the area a ghost town feeling. We also visited an abandoned RCMP outpost that operated from 1924 to 1951, where it is rumoured that one of the officers committed suicide due to the extreme isolation. We went for long hikes through the marshes and tundra of different ports on Baffin Island, at one point happening upon a musk ox enjoying an afternoon snack. Thirty-odd tourists snapped photos and giggled at his prehistoric hairstyle. We were not able to land on some sites due to the fear of polar bears running amuck, so we explored them from the zodiacs instead. Occasionally a polar bear popped its head up in curiosity, and it took everything we had not to jump up all at once and tip our boat!

Pond Inlet, located on the northern tip of Baffin Island, was an absolute delight. We met many of the 1,500 inhabitants. (I'm sure most of them came out to greet us since they don't receive visitors very often!) Exploring this northern Canadian community was fascinating. We visited their very informative museum, checked out the local hangouts, and quickly made friends with some of the local children. The manager's daughter even made my Pond Inlet hand-knitted hat, which I purchased in the museum, of course! Two members of our group had to play good cop/bad cop to get all the kids out of our zodiacs. I was ready to sneak a couple of them home with me – they were so sweet!

Lila and Madeleine were two young girls we met on our walks through the area, and they were more than thrilled to serve as tour guides. They talked about being on Christmas break

Ned & friends, Pond Inlet

and about their school, as their dog Snoopy followed along. We walked together for at least 45 minutes, down the street and onto the beach, where we learned they loved to pose for cameras. We had originally been shy about asking to take photographs but these two were like supermodels, with poses and grins for the camera. When we left them at the zodiac, they both insisted on signing my palms with their names, to which I replied, "I will never wash my hands again!" I waved goodbye until my newfound friends were long out of sight.

Our two days at sea were equally fascinating, giving people like me a chance to relax, take a zillion (god bless digital!) photos of the ocean and the vastness of the horizon, and (to be honest) take some well-deserved naps. Since we did not see sunset until Greenland, dinner often turned into evening drinks, which turned into plunge pool/sauna, which turned into 3:00 or 4:00 a.m. And with our wonderful daily wake up call from Aaron at 7:00 a.m. (I think everyone fell in love with his voice!) let's just say we all could use some rest. The seminars continued, with such topics as the Franklin Expedition history; Polar Bear 101 and 102; photography 101 (where I felt totally demoralized and my dreams of being a world famous photographer were shattered); icebreaker seminars and how to read nautical maps. There was certainly something for everyone.

Greenland was equally fascinating, from the brightly painted houses to the shy native peoples. There were a plethora of husky puppies around, making the girls squeal, "Oh, they're soooo cute!" which in turn made the guys laugh at our attempts to snap photos of them all. The hiking was wonderful, and kayaking around one of the biggest glaciers in the world, Sermeq Kujalleq in the Ilulissat Isfjord area, was amazing. The weather was in the low teens and sunny, making it a perfect afternoon. It only got better by the time our evening BBQ rolled around, with plenty of hot rum toddies to keep us warm.

Our last three days included visits to Ilulissat, Sisimiut and Kangerlussuag, again giving us the opportunity to meet the locals, stretch our legs, and continue to add delightful memories to the trip. It was not until our final morning that the melancholy struck - knowing this great holiday was coming to an end. This truly was, and I'm sure will continue to be, one of my favorite holidays. Being a nature lover certainly helped, but the diversity of the landscape, the uniqueness of the northern Canadian culture, the friendliness of the Greenlanders, and finding my sea legs all over again, made this a truly remarkable holiday.

From Sea-to-Shining-Sea: Across Canada with Kids

Mary Ellen Neilson (Grieb)

From the time that our children were babies, my husband and I decided that we wanted our kids to see every corner of Canada. The quest began in Ontario in the 90s with young toddlers, and finished in the Yukon in 2005 with teenagers. Since we wanted our children to meet their fellow Canadians firsthand, bed and breakfasts and university residences were the accommodations of choice. These accommodations also had the benefit of being easy on the pocketbook, allowing us to allocate limited travel dollars to special outings and activities. Here are some of the highlights from over a decade of travel in Canada.

British Columbia: We chose Salt Spring Island because it was rumored to be the second home of celebrities such as, Robin Williams and Barbara Streisand, plus it is one of the last vestige of hippiedom in Canada. Aside from wonderful beachcombing, craftsman tours, and the many organic farms on the island, our most memorable experience was to meet an authentic hippie family. The family was camping out at the sheep farm we were visiting, and our boys were amazed that this family sang at the farmers market and did odd jobs for the locals – talk about an alternative lifestyle! Although our kids noted the obvious benefits of no trips to the dentist and never attending school, this lifestyle offered maybe a little too much freedom, and we all agreed that we should keep our day jobs!

With Alberta as our home province, we had done the requisite trips into the Rockies many times, and so decided to explore the southern part of the province via RV rental. Alberta has over 200 provincial and recreational parks, allowing you to set up camp almost anywhere: in the Drumheller Badlands, the foothills of the Rockies, or the expansive prairie wheat land. The key with RV rentals is to pay attention to the "sewage seminar" that is given before you rent. Otherwise, you will have some very embarrassing and pungent experiences until you figure out how to empty the tank. In a real downside scenario, you will get the hoses mixed up (as we did) in which case you should stay at least 50 feet away from the people holding the hoses. You should also probably stay away from the hose people for at least 50 hours, or alternatively offer them a good supply of quarters, since they will need to stay in the campground and pay for showers for a very long time before the smell comes off! Aside from the sewage episode, the RV vacation was one of our favorites, as we were able to bring along both the dog and grandma, and had tons of family "together" time.

Saskatchewan: Home of the Royal Canadian Mounted Police and more than five million acres of parkland, we decided to focus our Saskatchewan travels on the tranquil and beautiful lake country located in the northern part of the province. Finding accommodations in small town Saskatchewan can be a challenge, as noted by our phone queries to the local

hotel in Good Soil. Our request for lodging prompted the response "You shouldn't stay here unless it's an emergency." This was followed by a referral to a golf course with a B & B who happily put up our brood for the night. Saskatchewan is where our kids learned the fine art of shuffleboard at the local community center, which is the center of activity in a small town. We also visited a buffalo farm and had front row seats at a forest fire, but our best memory is a hike that we took to a "historic site". After following numerous signposts – and hearing numerous complaints from our youngsters - we reached our destination. The historic site was nothing more than a buffalo skull on a signpost that read " On this site, July 24, 1673, nothing happened." No kidding.

Manitoba: Beautiful Manitoba, no wonder there is evidence of nomadic hunters here since approximately 10,000 BC… which is probably when the first mosquitoes started hunting here too! Surprisingly, our visit to Manitoba was during one of their drier summers, so we were able to miss the legendary mosquito season. Sans bugs, Manitoba as eastern most of the three-prairie provinces has numerous lakes and lots of culture in the capital city of Winnipeg. The three kilometres is a stretch of powdery white sand beaches along Lake Winnipeg are unbelievably Caribbean, with excellent kayaking through marshes amid choruses of bullfrogs. On the shores of Lake Winnipeg, in the village of Grand Beach, the locals know how to make their own fun with foot long hot dogs and an arcade set up in a local's garage. We were also very fortunate to be able to attend the Winnipeg Fringe Festival, although I'm not sure whether our kids were more impressed with the opportunity to attend their first improvisation comedy or to be the only kids in "a real bar" at the Winnipeg Free Press Club.

Ontario: Farm vacations and theme parks were the order of the day in Ontario when our children were small with Ontario as our home province. Ontario's vibrant rural life provided us with the most amazing trip. A stay at a dairy farm near "the other Woodstock " left us with memories of baby kittens and calves, plus the unforgettable chaos of milking time, where the cows seemed to think that my perspiring husband was the latest version of a salt lick. A later trip to Ottawa when the boys were preteens was considered a necessity, so that we could tour Parliament. You only need to sit on the front lawn of Parliament Hill to hear the 53-bell carillon ring from the Peace Tower. Apparently, bells and politics are far less important than chocolate, since our kids were more impressed with the trip to nearby Smiths Falls to tour the Hershey chocolate factory. We didn't unveil the secret of the Caramilk bar, but we were able to see Reese's Peanut Butter cups being made, with our kids wanting to "swim in the vats of chocolate." A factory outlet that sold 3 x 5 chocolate bars (feet, not inches) left us with stomach-aches for weeks and a supply of candy that lasted until the next Christmas.

Quebec: This was the province where we received the greatest resistance from our kids, since "they don't speak English and aren't really part of Canada anyway.. In our minds, this made the vacation even more important, and we chose a B & B in a working class Francophone district of east end Montreal just to make our point. We were the first westerners for the elderly couple at the B & B, and their numerous recommendations in halting English (supplemented by our limited understanding of French), led to a remarkable vacation and the beginnings of a wonderful friendship. A visit to a medieval feast and jousting match in St. Adolphe d'Howard – conducted entirely in French – was one of our more unusual evenings with our kids very quickly catching the drift to cheer for "le chevalier blanc!". The hometown performance by Cirque du Soleil in Quebec City was unforgettable, as was the visit to the magical shrine of St. Anne de Beaupre with its walls covered with crutches. Our very best memory, however, is from our whale-watching trip on a zodiac off the coast of Tadoussac. Our cold and wet children were getting tired of searching for the ever so elusive humpback, until we got covered in sea spray and made eye contact with an eye the size of a tire! The other important thing to mention from this trip is the genuine friendliness of the Quebec people, particularly in the small towns and rural areas. Although we had numerous language misunderstandings that sometimes led to interesting menu choices (how did we get onion rings and a side of honey anyway?), the trip left our children with the very clear understanding that Quebec is an important part of Canada.

New Brunswick: Everybody skips New Brunswick when they visit the Maritimes, which in our opinion is a big mistake since it was one of our favourite provinces. It is after all, the gateway to Atlantic Canada. The tidal phenomenon at Moncton is impressive, but New Brunswick offers so much more. Our stay at a farm near Sussex in the heart of dairy country resulted in an evening pondering the sunset, while our young boys pursued an aggressive bullfrog catch and release expedition at the pond located on the farm. Apparently, bullfrogs are excellent communicators with their gggrrrbits, since by sundown they were all clustered in the center of the pond out of the reach of our young fishermen! Our New Brunswick travels also included the island of Grand Manan, a little known spot located in the Bay of Fundy. Although the island seemed to be without telephones or TV's, the locals, who all bade us "gday," mesmerized our kids. Our tour of a restaurant housed in a school bus led to the obvious question from our boys as to why the restaurant was built in the bus. The answer? The bus had broken down and our hosts decided to flip burgers until they had enough for the repair bill! Best of all was the trip on a lobster boat four hours further out into the Bay of Fundy. This led to an island that was a main puffin migration stopover and mating ground. Thousands of shy puffins were an amazing sight to see, as were hundreds of seals on a neighboring island. Our nature viewing was somewhat impeded, however, by a large troop of self-appointed island defenders – the dive-bombing kamikaze arctic terns that seemed to have it in for our kids!

Nova Scotia: No visit to Nova Scotia is complete without a tour of the Cabot Trail, which is one of the most scenic drives we have ever encountered in the world. The trail winds around the rocky north shore, and then ascends to the incredible plateaus of Cape Breton Highlands. We felt the beauty of the Highlands was best captured by the words of its most famous son, Alexander Graham Bell. He once said, " I have travelled around the globe. I have seen the Canadian and the American Rockies, the Andes, the Alps and the Highlands of Scotland's, but for simple beauty, Cape Breton outrivals them all." The province is also rich in tradition and folklore, so no visit to Nova Scotia is complete without a close encounter with a Cape Bretoner. Cape Bretoners have many stories, and they are usually long, so be prepared to get seriously off schedule. Our accommodations in Cape Breton ranged from a room above a barn on a farm of Icelandic horses, to an abandoned convent that housed a talkative old fisherman and his nocturnal cat – with the cat causing a lengthy round of allergic sneezing by our son when it crept up onto the pillow (it was the cat's room after all). The living history of Fort Louisburg and the Miners Museum in Glace Bay were of course very memorable, but our children still speak of the town of over 100 scarecrows, near Margaree Harbour. Every year the townsfolk add another scarecrow scene, and accompany it all by an eerie background of Celtic music. It was pretty neat viewing in the safety of broad daylight for the kids, but a little too much in the style of Stephen King to visit at night.

PEI: This is the island that inspired Lucy Maud Montgomery to write the internationally famous books of *Anne of Green Gables,* and as everyone says, Prince Edward Island is cute. There are cute roads, cute theme parks, and cute lobster suppers at the local church halls. Even the massive Confederation Bridge makes a rather cute backdrop for family travel photos. Not so cute are the purple jellyfish that hang around the Northumberland Strait, which the locals nonchalantly toss aside during a swim to avoid painful stings (we decided to wade instead). Also not cute were the highly serious PEI "border guards" at the ferry docks adjacent to Newfoundland. Based on their dour expressions – and the thoroughness of their rummage expedition through our car trunk - we figured we had become embroiled in a major Newfie drug bust. Not! In PEI nothing is more important than potatoes, and our searchers were in fact looking for smuggled spuds carrying blight or other diseases. We never did find out the penalty if we had dared undertake a spud smuggle – maybe they would have shipped us back to "The Rock?"

Newfoundland: The 13 hour ferry ride to Newfoundland was a real challenge for our family. My husband had cracked his ribs on a water slide just before the trip, so every hurl of the boat elicited a moan. It also turned out that our youngest son was not seaworthy, so every hurl of the boat led to more hurls from him. Add to the mix an open dormitory with endless rows of bunk beds (populated by other hurlers and moaners), plus an onboard pub next door (with the requisite ballad singers and all-night revelers) and it made for a very long night. We had chosen the western side of Newfoundland for our trip, in order to hike Gros Morne Park and the Tablelands. Gros Morne Park represents some of the most incredible scenery in Canada, with plenty of campgrounds and hiking trails to enjoy. We were looking forward to viewing the Tablelands, lunar like landscapes with ancient volcanic formations. The exploration of the area is world-renowned and would definitely be the highlight of any trip to Newfoundland. Unfortunately, hiking was now out of the question in view of the family injury list. Instead we holed up in the rented cabin, watching the Twilight Zone on our single TV channel, and listening to our next-door neighbor chain saw wood at 6:00 a.m. Our best memory of Newfoundland is the people, who were wonderfully friendly, although the language barrier was almost as challenging as Quebec. For example "How many yars be the byes" was actually a question about the ages of our two boys.

Yukon: Canada's True North. The word "Yukon" means "the Great River" or "big river." Although technically the Yukon is only one of the northern territories, we agreed it was suitably representative given the remote possibility of ever getting our teens to visit all three territories. Our visit to the Yukon was in the fall, since the primary reason for our trip was to experience the Aurora Borealis. (Aurora Borealis, or northern lights are caused by huge explosions on the surface of the sun that sends out streams of charged particles that interact with the earth's upper atmosphere). What a hopeless mission that turned out to be: every night we set our alarms for 1:30 a.m., only to find we had either just missed the northern lights or they hadn't come out yet. Weary of this nightly drill, we rationalized that natural phenomenon were vastly overrated anyway, and called it quits. The very next evening, during our usual midnight soak in the hot tub, out came the Aurora on the eve of my husband's birthday, performing a spectacular crackling light show across the sky. The northern lights lit up the sky like a sheet of magical ribbons gently travelling waving across the sky, leaving our cynical teens speechless. We found it ironic that when we stopped chasing the lights, they came to us – which kind of epitomizes the North.

I have nothing to declare
except my genius.

Oscar Wilde at New York Customs

Purchase Compass Companions for a friend

Compass Companions: *A Collection of Canadian Travel Stories* ______ x \$23.95 = \$______
Shipping & handling charge (total order) \$5.00 for 1-4 books ____________ = \$______

Order five or more books and receive free shipping
(applies only if mailed to same address in Canada) Subtotal ________ = \$______

In Canada add 7% GST or 15% HST where applicable. ____________ = \$______

U.S. and international orders payable in U.S. Funds. Price is subject to change.

Please Print

Name ____________________

Address ____________________

City __________ Province __________

Postal Code __________ Phone __________

E-mail __________ Facsimile __________

Please make cheque or money order payable to:
Compass Companions
Silver Springs P .O. Box 71094
Calgary, Alberta T3B 5K2 Canada

For school fundraising or unit purchase prices, please contact Compass Companions at www.compasscompanions.ca

Please allow 2-3 weeks for delivery.

Order online at www.compasscompanions.ca

EPILOGUE

From the time I flew in my first airplane – a small single engine type flown by a friend of my father (the only doctor in Fort McMurray at the time) – from Edmonton to Fort McMurray, I was hooked on travel! We landed on a strip of grass and I recall looking up at the tall narrow trees on either side of the "runway" stretching as high as the sky. I saw the big picture… that was fun!

I could not believe my luck when I met Larry, my husband, who had the same travel passion I did. We would stay up for hours viewing slide shows of Africa, India, Thailand and many many other destinations. Then we would have our friends over, view their slide shows and share in the excitement of the adventures. Another lucky break, I have ten wonderful years working for a airline whose corporate head office was based in Calgary. Weekends in Vancouver and Hawaii were our frequent way to see the Pacific Ocean. Now we travel with our children, seeing old familiar cultures and new ones through their eyes and enjoying the newness of it all.

Travel has become too familiar for so many as we can instantly access readily available guides and destination data. Yet, I prefer the "old shoe" way of travel, the more you wear it, the more comfortable it becomes. It's not a matter of where you go but what you do when you get there. This is when I came up with the idea to share in the comfort of your home, the unique experiences that Canadian travellers encounter around the world. So began this book *Compass Companions*.

We live in Calgary, Alberta, Canada, backyard to the Rocky Mountains!